TERRESTRIAL ENVIRONMENTS

J.L. CLOUDSLEY-THOMPSON

CROOM HELM LONDON

Croom Helm Ltd., 2-10 St. John's Road, London SW11

ISBN: 0-85664-001-8 hardback
0-85664-180-4 paperback

Printed in Great Britain by
Redwood Burn Limited
Trowbridge & Esher

CONTENTS

PREFACE

In this volume are described the more important terrestrial habitats of the earth, with special respect to their influence on the fauna. Climate and vegetation are outlined only briefly in so far as they affect animal life. Finally, I have tried to illustrate the selective influence of the environment on its fauna and, conversely, the influence of animals on their habitat.

There are few regions of the globe which have not been altered to some extent by human activity. Indeed, of the various biomes described, only tropical forest, taiga, tundra and snowlands remain in anything like their pristine state. Here I have been concerned with climax vegetation and natural faunas, despite the fact that both have been greatly degraded by man, the self-styled 'cultured primate', whose population explosion threatens the ecological stability of our planet.

The greatest manifestations of animal activity are those concerned with locomotion, feeding, escape from enemies and reproduction. Structural adaptations are correspondingly related to the spatial level at which an animal lives – under or on the surface of the earth, in trees or in the air. The dominant fauna of a region is therefore dependent to a large extent upon the vegetation; forest, grassland, desert or tundra.

A second class of adaptations is concerned with obtaining food. These again are affected, directly or indirectly, by the vegetation and may involve weapons of offence and defence, sense organs, colouration etc. – characters which are also related to escape from enemies. Finally, a smaller number of adaptations is concerned with the protection and development of the offspring. These adaptations are less closely related to the environment than are the others mentioned above. Where appropriate, however, all three are discussed in the following chapters in relation to the environments inhabited by their possessors.

In each kind of country, therefore, certain forms of animal life predominate and possess adaptations in common; others are characteristically rare or absent. Animals are thus moulded and adapted to their environments by structure, behaviour and life history. But they, in turn, modify those environments by feeding on the vegetation, destroying trees, overgrazing, causing soil erosion, damning streams, creating swamps and, on a subtler but more significant level, by producing humus, affecting the moisture-holding capacity of the soil, and thereby determining the vegetation that it supports. In this way, the fauna, flora and physical environment each act upon, and are acted on, by the others. Selection pressures are consequently generated and these then engender new genetic patterns.

The vast complexity of interactions within and between the various ecosystems of the earth make it impossible in any work of man to 'hold the mirror up to nature'. But the attempt is not without its own rewards.

> Thrice happy he who, not mistook,
> Hath read in Nature's mystic book.
>
> Andrew Marvell (1621-78)

ACKNOWLEDGEMENTS

My warmest thanks are due, as always, to my dear wife Anne for her advice and help, which included drawing Fig. 19, and to Margaret Sanderson for typing the manuscript.

1 ZOOGEOGRAPHY

The way in which plants and animals are distributed in the world results from a combination of factors. These include soil, climate and past history. Since life first began, the surface of the earth has undergone immense changes. Until perhaps 200-500 million years B.P.,* the continents were probably grouped close together, with an ice-cap at times covering parts of Africa and much of South America, India and Australia. Since then, they have drifted apart, India moving thousands of miles to the north-east until it collided with Asia. The separation of the continents isolated large groups of animals and is responsible for many of the peculiarities of the faunas of different regions of the world.

Continental Drift

The theory of continental drift was proposed by Wegener (1924) to explain some of the stranger facts of animal distribution, previously accounted for by an hypothesis of land bridges (Suess, 1904). For a long time it had been thought that, until the beginning of the Cenozoic era, the land mass of the world had been divided into two halves – a northern part comprising Europe, northern Asia, Greenland and North America; and a southern Gondwanaland, which included Africa, South America, Australia, New Zealand and Antarctica (Fig.1). These two super-continents were separated by an extensive Tethys Sea, of which the Mediterranean still persists as a remnant.

The continuity of the southern land was thought to account for the distribution of animals such as side-neck turtles (Chelyidae and Pelomedusidae), flightless ratite birds and marsupial mammals, which are found only in southern continents today. South America was believed to have been joined to Africa by a South Atlantic bridge (Atlantis). A second land bridge was the supposed continent Lemuria which was said to lie between Africa, the island of Madagascar, and India; while a third bridge was thought to have connected South America to Australia through Antarctica. There is no geological evidence for the existence of such land bridges and, although these were originally postulated to explain biological anomalies, in fact the whole idea raises more problems than it solves. The subject has been discussed by George (1962). It is the consensus of opinion today that the theory of

* B.P. – Before the present

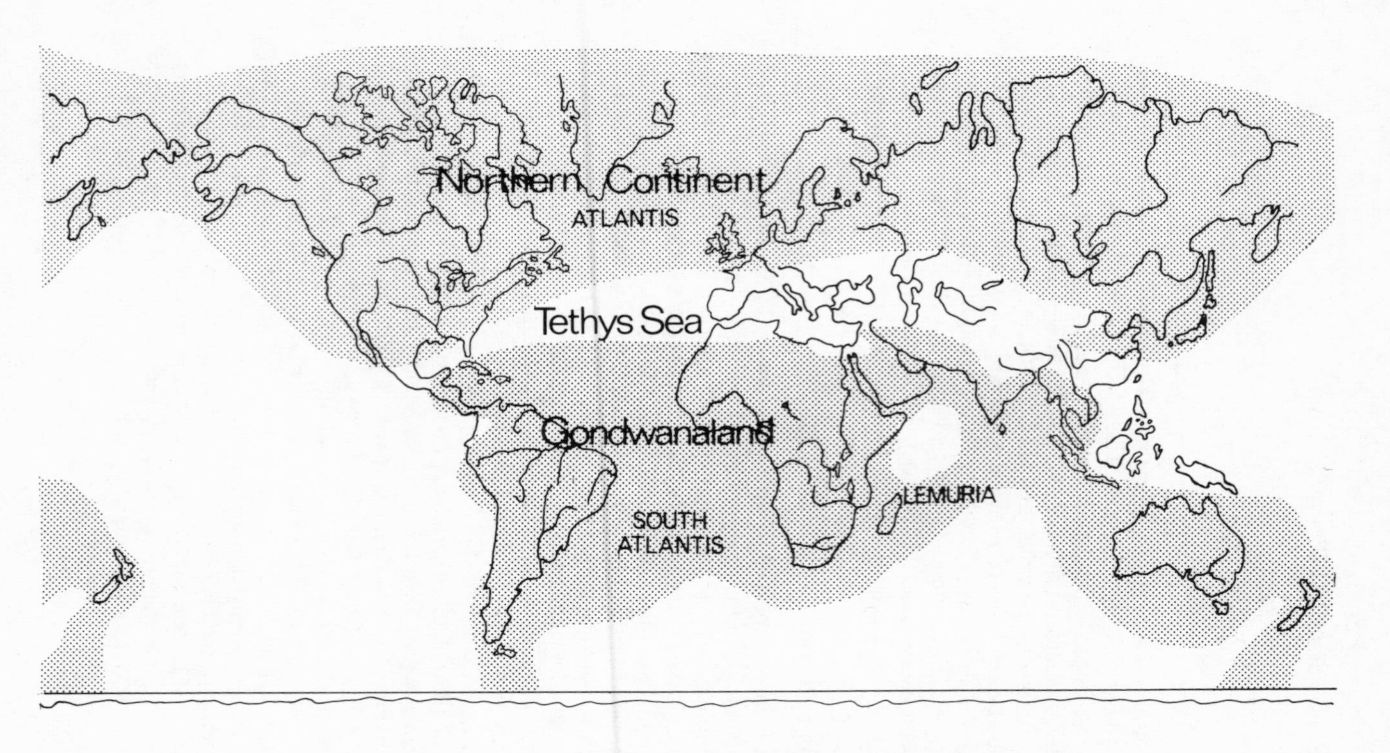

Fig. 1 The world, showing hypothetical land bridges.

continental drift may provide a better explanation of biological facts than does the hypothesis of land bridges or that of the permanence of the continents.

Wegener (1924) suggested that all the continents had been joined together in a single land mass until after the Carboniferous period, when they began to break up and drift apart. He also supposed that, when they were joined together, they would have shared the same fauna. This theory would equally well explain the presence today of flightless birds, side-necked turtles and lung fishes in all three southern continents. Later on, Australia and Antarctica became separated from Africa, except through South America (Fig. 2), and the fauna of the southern continents would have become differentiated into two main components. Australia and South America shared the marsupials, hylid tree frogs and so on, while South America and Africa shared groups such as ostriches, pipid toads and characin fishes.

Wegener, and later supporters of his theory, envisaged the clustering of the southern continents around Antarctica, either as part of a single world continent, Pangaea, or of a southern super-continent, Gondwanaland. They based their theory on four major lines of geological and fossil evidence: (a) The fact that the contours of the continents can be fitted together reasonably neatly. (b) Important geological formations in the various continents are of similar age or duplicate each other structurally. (c) Africa, South America, Australia, India and Antarctica shared a widespread glaciation in the southern hemisphere in Permian and Carboniferous times. (d) During the Permian, southern continents also shared a unique flora, dominated by the genera *Glossopteris, Gangamopteris, Neoggeranthiopis* and *Paracalamites* (Keast, 1971).

Additional evidence is afforded by the distribution in southern continents today of earthworms of the genus *Microscolex,* Onychophora, woodlice (Styloniscoidea), spirostreptid millipedes, various harvest spiders (Opiliones), spiders, Ricinulei, Collembola, insects, cichlid fishes, amphibians (Leptodactylidae), reptiles, birds and mammals, both recent and fossil. The study of parasites is also useful in the interpretation of the distribution of the southern faunas. Examples are afforded by nematodes and trematodes, as well as by the Mallophaga of the African ostrich and South American rheas. These belong to the same genus *(Struthiolipeurus).* (For additional examples, details and bibliography, see Legendre and Cassagne-Mejean, 1967-68.)

Vigorous arguments were raised against the theory of continental drift in Wegener's day and subsequently, for there was no known mechanism by which continents could move. Then Irving and Green (1957) brought new light to the problem by their research on rock magnetism. In some rocks, probably not more than 5 per cent of those exposed on the earth's surface, the direction of magnetization can be

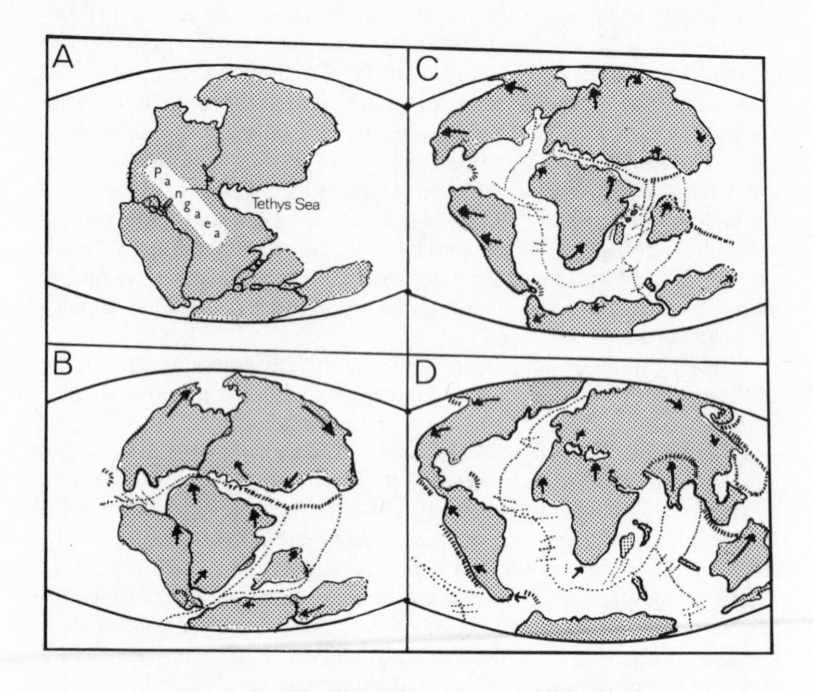

Fig. 2 Continental movements during the Mesozoic.
Reconstruction according to Dietz and Holden (1970) and based on the latest marine paleomagnetic and sea-floor spreading data and ideas. These authors assume the existence of a single supercontinent (Pangaea) in the Permian (A) and thus differ from some other authors who postulate both southern and northern continents. The southern and northern continents, Gondwana and Laurasia, separated by the Tethys Sea, appeared at the end of the Triassic (B), by which time South America, Afric and India had broken free from Antarctica. By the end of the Cretaceous (C), South America and Africa had drifted well apart, and the Indian plate was approaching Asia. According to this view, Australia did not separate from Antarctica until the end of the Cretaceous. The Cenozoic distribution of the continents is shown in D. Rifts and oceanic ridges, from which sea-floor spreading is occurring, are shown by dotted lines, subduction zones by the banded lines. (After Keast, 1971).

identified with the direction of the geomagnetic field at the time of their deposition. A study of the newer volcanics of Victoria, Australia, provided the first confirmation from the southern hemisphere of the average dipole nature of the geomagnetic field during the later part of the Cenozoic. This observation necessitated resurrection of the idea of large-scale translational and/or rotational movements in Cretaceous and Tertiary times, for such a conclusion is contingent on the dipole hypothesis and cannot be avoided if it is true. Palaeomagnetic results from the Permian period of Africa likewise suggest that this continent occupied a different latitude at that time (Gough, Opdyke and McElhinny, 1964). They also support the hypothesis that Africa then lay adjacent to Antarctica. The plotting of Palaeozoic palaeomagnetic pole positions indicates a former clustering of the southern continents and of India in the far south (for references, see Keast, 1971); and work in Australia suggests a series of changes in the palaeolatitudes of that continent, indicating a progressive drift northward across the Southern Ocean (Irving, Robertson and Stott, 1963).

Various dates have been advanced for the break-up of Gondwanaland. These range from the end of the Cambrian or the end of the Permian to late Triassic, or even mid-Jurassic period (Keast, 1971). Dietz and Holden (1970 a, b) suggested that the break-up might have begun in the Triassic and occurred in a series of stages (Fig. 2). Thus, while the split between the South American and African continents possibly began in the early Mesozoic era, simultaneously with their separation from Antarctica, it may not have been finally completed until the Cretaceous: and even later dates have been suggested by other authors (Keast, 1971).

Comparisons of the palaeomagnetism of terrestrial rocks serve as indicators of the relative displacement of the continents during geological time, but they can be interpreted in several different ways. Moreover, they provide information only on palaeolatitudes, and not on palaeolongitudes. The theory of plate tectonics, however, has recently been proposed as a mechanism for continental drift (Dietz and Holden, 1970a, 1971; Keast, 1971). Despite the fact that its details have not yet been clarified, this theory has removed most of the original objections to the concept of continental drift. It stems from recent findings with respect to bathymetry, volcanicity and earthquakes activity, as well as from the analysis of deep-sea cores and of anomalies in the earth's magnetic field over the oceans.

The earth's crust is now envisaged as being composed of a rigid outer shell or lithosphere, perhaps 100 km thick, resting on a less rigid asthenosphere. The lithosphere is broken up into six major plates and some smaller sub-plates that are in a state of motion relative to each other. These contain both continental and oceanic segments of the earth's crust, but are rigid and more or less free from internal distortions.

In terms of plate tectonics, continental drift involves the break-up of super-continents followed by passive movements of fragments of continental crust across the surface of the earth. How these movements take place is not yet known: nevertheless, the reality of plate tectonics is now clearly established.

Although it is generally accepted that a southern continent, Gondwanaland, persisted until about the mid-Triassic, or even later (Cracraft, 1973), there is less agreement about the timing and pattern of the phases of its subsequent break-up. While geologists may concur that the South Atlantic, for example, opened up in the Jurassic or early Cretaceous period, the data derived from the various techniques employed are conflicting when applied to, say, the history of India (Keast, 1971). Perhaps the most durable reconstruction has been that of du Toit (1937). This was based on detailed geological information, is similar to the geometric fit, and is generally consistent with palaeomagnetic data. The positioning of Madagascar is critical in most postulated reconstructions of eastern Gondwanaland. Although du Toit, and other authors, have suggested a northerly derivation, Tarling (1972) believes that there is good geological and geophysical evidence in support of a southerly origin. This would place Madagascar against Mozambique, with India close by and Antarctica separating Australia and New Zealand from the other southern continents.

Zoogeographical Realms

The world was first divided into zoogeographical regions on the evidence of its bird fauna. In 1858, Sclater named six avifaunal regions which were only slightly modified when, in a monumental work, Wallace (1876) extended this concept to include all the land animals, both vertebrate and invertebrate, whose ranges were then known. The six zoogeographical realms are as follows: Palaearctic, Ethiopian and Australian, forming the Old World (Palaeogaea), and the Nearctic, Oriental and Neotropical which comprise the New World (Neogaea) (Fig. 3). Other early classifications are discussed by Evans (1899).

During the last million years or so, the northern temperate regions have several times been locked in the grip of ice ages which caused the extinction of many species of animal that had insufficient time to emigrate or to become adapted to changing conditions. During the last ice age, for example, moles *(Talpa europaea)* died out throughout western Europe except Spain. When the ice retreated, however, moles moved back northwards, reaching England before the Channel formed; but their route, like that of snakes, was blocked by the Irish Sea. Ice also provided a route for other species to travel across frozen seas, and also gave rise to land links by reducing sea levels. One such link was a

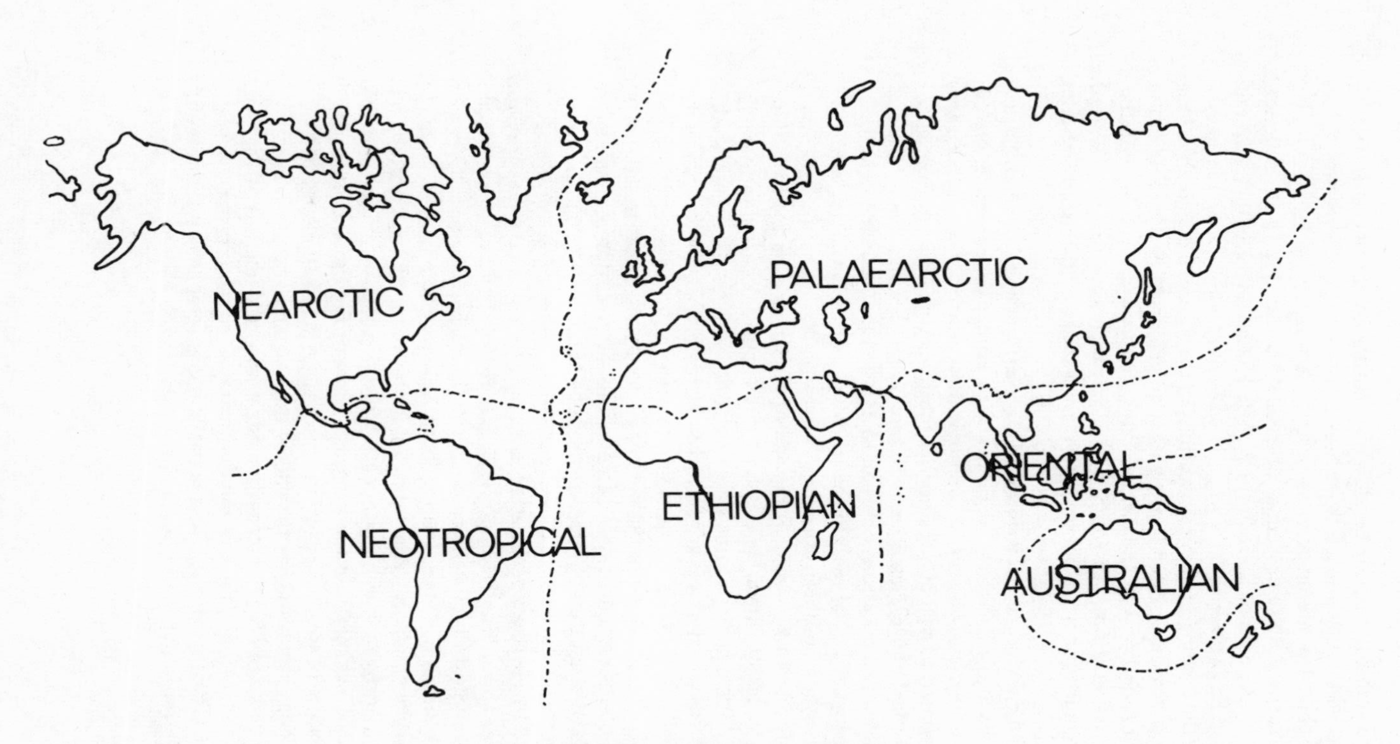

Fig. 3 Zoological regions of the world.

bridge across the Bering Strait. This was used by the ancestors of modern horses when they crossed from North America into the plains of Eurasia more than 1.5 million years ago.

Likewise, the camel family originally flourished in North America but became extinct there in the Middle Pleistocene, less than two million years B.P. The sub-order Tylopoda was unknown in other continents until the coming of desert conditions in the Pliocene epoch about ten million years ago, when forms with digitigrade and padded feet entered South America, Asia and Africa. Tapirs *(Tapirus* spp.*)* are now found only in South America and S.W. Asia but their ancestors were widespread in the northern hemisphere when there was dry land across the Bering Strait.

The Pleistocene glaciations had three effects upon the distribution of animals in the Palaearctic region. The spreading ice, of course, erased life during its advance; it drove the various zones of life southwards; it engendered pluvial periods further south in Africa. In addition, as mentioned above, glaciation withdrew so much water from the sea as to create land connections between islands and the mainland, and to affect the growth of coral reefs in the tropics (Elton, 1958).

The three southern continents were supplied with different mammalian raw material at the end of the Cretaceous and were then more or less isolated. Major faunal interchanges occurred only rarely – with Africa, perhaps three times, and once with South America. All other colonisation was adventitious. The latest palaeontological data strongly support the contention that each continent received its mammalian fauna from the north and not by any southern connection, since continental drift took place before the evolution of mammals.

The distribution of micro-arthropods in the South Polar region supports the conclusion that a relic fauna is present, which has been joined by elements penetrating from the sub-Antarctic zone (Wallwork, 1973). The alternative hypothesis that distribution can be explained in terms of post-Pleistocene colonisation is far less probable.

At the time of its isolation, South America had a fauna restricted to marsupial carnivores, edentates and primitive ungulates: Australia had only marsupials and monotremes. Africa presumably obtained a faunistic nucleus from the Pangaean fauna of the Cretaceous period, but the nature of this nucleus is unknown since the fossil record of the continent begins only at the end of the Eocene (Keast, 1968a). Sidorowicz (1971) has recently made a statistical analysis of the occurrence of members of the Carnivora which permits the characterisation of the fauna in particular geographical provinces and the definition of their boundaries.

Palaearctic

The Palaearctic realm is the northern part of the Old World. It includes the whole of Europe and Asia, except for that portion lying south of the northern regions of Iran and Afghanistan, the Himalayas and the Nan-Ling mountains of China, all of which form physical barriers which most animals do not cross. Likewise, the Sahara now creates a barrier to animal migration. The Palaearctic also includes the Arctic islands to the north of Siberia and the northern parts of Africa, Iceland, Spitzbergen.

The vertebrate fauna of the Palaearctic realm is not very rich. Its characteristics can be summarised as a complex of Old World tropical and New World temperate families. For example, of 28 families of Palaearctic land mammals, excluding bats, nearly one-third is widely distributed. This includes the hares and rabbits (Leporidae),* mice, voles etc. (Muridae) and dogs (Canidae), which are world-wide in their distribution, and the shrews (Soricidae), squirrels (Sciuridae), hamsters (Cricetidae), badgers, otters, weasels and stoats (Mustelidae) and cats (Felidae) which are found in every realm except the Australian. The Palaearctic shares the bears (Ursidae) and deer (Cervidae) with Nearctic, Neotropical and Oriental regions; the buffaloes and wild cattle (Bovidae) with the Nearctic, Ethiopian and Oriental realms; and the hedgehogs (Erinaceidae), Old World porcupines (Histricidae), mongooses and genets (Viverridae), hyaenas (Hyaenidae) and pigs (Suidae) with the Ethiopian and Oriental. In contrast, the racoons (Procyonidae) are shared with the New World; dormice (Gliridae), jerboas (Dipodidae), hyraxes (Procaviidae) and horses, asses and zebras (Equidae) with the Ethiopian, and moles (Talpidae), pikas (Ochotonidae), beavers (Castoridae) and jumping mice (Zapodidae) with the Nearctic region.

Only two families of mammals, the mole-rats (Spalacidae) and the rare desert dormouse (Seleviniidae) are restricted to the Palaearctic realm (George, 1962). Similarly, almost all the birds of the Palaearctic belong to families having a very wide distribution. The reptilian fauna of the realm is poor, but there are many tailed Amphibia, as well as Anura.

Nearctic

The Nearctic region includes North America, from Greenland in the east to the Aleutian islands in the west, and extends as far south as the southern limits of the Mexican plateau. Except for the narrow strip of Central America, it is now separated by sea from all other geographical realms.

---

* In the interest of precision I have used scientific terminology throughout: and, when English names are given, I have nearly always added the Latin equivalents.

There are even fewer families of mammals in the Nearctic than there are in the Palaearctic, and many are unexpectedly absent. Of four endemic families, three are rodents and one is an artiodactyl. This family is represented by a single species, the pronghorn *(Antilocapra americana)*, an interesting form related both to the deer and to the Bovidae but differing from both in the construction of the short, branched horns of the males. These horns have a bony core: as in deer, there is a soft covering, but this only is shed annually, not the entire structure.

Nearctic birds are even less differentiated from neighbouring regions than are the mammals, but the reptile fauna is extremely rich and has more endemic families than that of the Palaearctic realm.

Neotropical

The Neotropical region includes South America, the West Indies and most of Mexico. Just as the Ethiopian realm is separated from the Palaearctic by the Sahara, so the Neotropical is isolated by desert from the Nearctic. The fauna of the Neotropical realm is both distinctive and varied. Excluding bats, 32 families of mammals are represented in the region, of which sixteen are unique. This is the highest number of endemic mammalian families found in any zoogeographical realm. Three of them, together, form the order Edentata: this contains the anteaters, sloths and armadillos, and is almost confined to the Neotropical. Only one species, the nine-banded armadillo *(Dasypus novemcinctus)*, has successfully colonised the Nearctic and extends through Mexico into Texas.

The remaining endemic families of Neotropical mammals include the rat-like Caenolestidae, one of the two families of marsupials that occur in the region, two families of monkeys and eleven of caviomorph rodents; not to mention five families of bats, including the vampires (Desmodontidae). The South American bird fauna is equally striking. Two orders, Rheiformes and Tinamiformes, are confined to the realm as, indeed, are nearly half of the remaining families of birds – including toucans (Ramphastidae), trumpeters (Psophiidae) and the hoatzin *(Opisthocomus hoazin)*, sole representative of the Opisthocomidae.

Reptiles abound: one family of side-necked turtles (Pelomedusidae) is shared with Africa, another (Chelyidae) with Australia. Amphibians are nearly all frogs and toads (Anura) in marked contrast to the fauna of the two northern continents. Only a couple of genera of urodeles or salamanders *(Oedipina* and *Bolitoglossa)* are widespread, but a few axolotls have entered Mexico from the Nearctic. Thus the Neotropical realm is rich in endemic families of vertebrates of all classes and, of the more widely distributed families, it shares many with the nearctic and several with other tropical regions of the world (George, 1962).

The Middle American province of the Brazilian sub-region was the primary centre of origin, evolution and dispersal for the mammals of South America. Faunal interchange with the Patagonian and West Indian sub-regions must have taken place since the Middle Tertiary at the latest, according to Hershkovitz (1969). Of the three southern continents, South America was more isolated than Africa, but less so than Australia, before it was colonised by North American mammals (Patterson and Pascual, 1968). The date at which this took place is still a matter of dispute (Keast, 1969). It was followed by a spectacular extinction, mainly of large mammals, of both northern and southern ancestry, at the end of the Pleistocene. The arrival of man was most probably the cause of this (Patterson and Pascual, 1968).

Ethiopian

The Ethiopian region includes Africa south of the Sahara and the southern part of the Arabian peninsula. Like the Neotropical realm, it has been isolated by sea since the dispersal of Gondwanaland, but is now connected by land with its northern neighbours. In its mammalian fauna, this is the most varied of all the zoogeographical realms and contains 38 families excluding bats (Chiroptera): in number of unique families, it is second only to the Neotropical, for there are twelve. These include the giraffes (Giraffidae), hippos (Hippopotamidae) and aardvarks (Tubulidentata), three families of Insectivora and six of rodents

Birds are numerous in the Ethiopian and, like the mammals, have strong affinities with the Oriental region: there are only six exclusive families. Reptiles, amphibians and fishes, in contrast, show greater affinities with the fauna of the Neotropical realm (George, 1962). Africa was in the centre of Gondwanaland so it is not surprising that the older and less mobile components of its fauna should be related more closely to those of the other southern continents while the avifauna and mammals show closer affinities with the Oriental and Palaearctic. The influence of Gondwanaland on the fauna of Africa is discussed in detail by Jeannel (1961).

Oriental

The Oriental region includes India, S.E. Asia, southern China and the western isles of the Malay Archipelago. It is demarcated from the Australian realm on the south-east by Wallace's Line which separates Borneo and Bali from Celebes and Lombok. Wallace's Line was so named by T.H. Huxley in recognition of the fact that Wallace (1860) had first pointed out that an abrupt change in the pattern of the fauna occurs when this line is crossed (Elton, 1958).

Although the fauna resembles that of the Ethiopian realm, it is neither so rich in endemic families nor does it exhibit the same variety of widespread ones. Of 30 mammalian families, excluding Chiroptera, only four are endemic. Birds and reptiles are well represented, as are Anura, but only a few urodeles from the Palaearctic have reached the northernmost parts of the region (George, 1962).

Australian

In addition to Australia, this region embraces Tasmania, New Guinea and islands of the Malay Archipelago east of Wallace's Line. New Zealand and the islands of the Pacific are not included. This region is unique in having no land connections with any other region since the break up of Pangaea.

Some years after Wallace's Line had been recognised, another line was chosen to separate the Oriental and Australian regions. This was called Weber's Line. It runs between the Moluccas and Celebes, and between the Kei Islands and Timor, and was based mainly on observations of the mammal and molluscan faunas of the area, whereas Wallace based his line mostly on birds. Geologically, Wallace's Line marks off the eastern limit of what was once a land mass joined to Malaya, while Weber's Line more or less indicates the westerly limit of the Australian continent. The islands in between these lines probably lay beneath the sea during most of the Cenozoic period. When they re-emerged, they were colonised by haphazard migrations across the sea from both the east and from the west.

The paucity of the vertebrate fauna of the Australian region is its most striking characteristic. Apart from bats, there are only nine families of mammals, but eight of these are unique. The dominant mammals are marsupials, except for recent introductions from the Palaearctic of rabbits, foxes, rats and murid mice. Dingoes and pigs were brought in during prehistoric times.

The absence of native placental mammals, except bats, indicates that the region has been isolated from Asia since early Tertiary times. Since there was no competition from placentals, the marsupials were able to evolve luxuriantly and radiate into various ecological niches occupied in other realms by placental mammals. This parallel evolution is not complete, however, since there are no marsupial bats, seals, or whales, and placental representatives of the orders occupy the air and seas of Australasia. Nevertheless, virtually all major ways of life except those of rodents and bats, have been filled by marsupials, whose great diversity of body forms reflects this.

Convergence with placental counterparts on other continents may involve the form of the entire body, but is usually restricted to single structures or groups of structures. Within the limits of the genetic potential, the closeness of the similarity must, in part, reflect the vigourousness of the selective pressures associated with entering that particular way of life. Thus the mole-like body approaches the ideal for its role and has been evolved independently many times in mammalian evolution, as well as in *Notorcyctes typhlops.* On the other hand, larger terrestrial grazers are not so limited and kangaroos, accordingly, are convergent with ungulates only in their lengthened muzzles, dentition, and alimentary structures. At the other extreme, generalised arboreal herbivores such as oppossums *(Trichosaurus* spp.*)* and carnivores *(Dasyurus* spp.*)* show few truly convergent trends toward ecological equivalents among the placental mammals. For each organism is, in effect, a mosaic of semi-independent characters (Keast, 1968b).

Two Australian families comprise the only living Monotremata – the spiny ant-eaters (Echidnidae) and the duck-billed platypus *(Ornithorhynchus anatinus).* The egg-laying mammals are only distantly related to the marsupials and placentals: they may even have had a separate origin from the Reptilia.

The avifauna of Australasia does not rival that of the mammals in peculiarity, for most species belong to families that have a wide range. Nevertheless, ten families are unique and include two of flightless birds, emus (Dromaiidae) and cassowaries (Casuariidae). Reptiles, too, are only moderately varied, with two exclusive families and sharing side-necked turtles with the Neotropical region. Amphibians are few, there are no urodeles and the freshwater fish fauna is equally poor.

Part of the Australian vertebrate fauna – frogs, turtles and marsupials – resembles that of South America. Another part, comprising terrestrial reptiles, many birds, and placental mammals, shows close affinities with the Oriental region: there is little in common with the Ethiopian (George, 1962).

Island faunas

Continental islands tend to have a reduced version of the mainland fauna of the zoogeographical realm to which they belong. Oceanic islands such as St. Helena and the Galapagos, on the other hand, have very poor, sparse faunas. Often, a group of closely related species will have radiated to fill a number of ecological niches. A well-known example is afforded by Darwin's finches which have become adapted to feeding on various types of diet. The chief morphological variation lies in the bill. This may be adapted for eating seeds; long and pointed for eating cactus; parrot-like; or insectivorous. *Camarhynchus pallidus* not only excavates into wood, like nuthatches *(Sitta* spp.*)* but has the unique habit of picking up a cactus spine, or breaking off a small twig, and probing with it into crevices from which insects are driven out (Lack,1947).

Like the Galapagos Islands, New Zealand and Madagascar have long been isolated, probably since the Mesozoic, and have received their respective faunas from across the sea by chance introductions which have subsequently radiated to fill vacant ecological niches.

2 ENVIRONMENTAL FACTORS

The world is populated by living organisms which can survive only under particular environmental conditions. These include a number of chemicals that will allow metabolism, reproduction and variation to take place. Of the 96 elements that are believed to constitute the universe, at least 36 and probably as many as 46 are major or minor constituents of protoplasm (Allee *et al.,* 1949).

Since life is based principally upon compounds of carbon, hydrogen, oxygen and nitrogen, organised in extreme complexity, the temperatures at which it can exist are limited. The high temperatures of the stars would exclude the possibility of any organisation of molecules sufficiently complex to serve as a basis for life; while extreme cold slows down chemical processes so much that life is impossible near absolute zero. Although living protoplasm, in its latent stages, may survive at temperatures as low as -270°C or as high as 150°C, life is limited in practice to a much narrower range.

Finally, life can exist only within a limited range of density and pressures. Molecular organisation would be impossible under extreme pressure while, at very low pressures, molecules could not assemble and align the chemical units necessary for life. From this, it follows that a viscous state, intermediate between an ideal solid and a liquid, is needed to support the chemistry of life. Such is provided by the colloid state, Between gel and sol there is sufficient solidity to permit organisation, and enough liquidity to allow change.

Living organisms are dependent upon sources of energy that can be liberated during metabolism. These basic sources of energy are provided by plants, which are able to synthesise carbohydrates and proteins. In addition to supplying food for animals, plants also provide them with shelter and protection. Consequently the fauna of a region is dependent upon the vegetation. This, in turn, depends upon two interrelated environmental factors – climate and soil. The biology of the environment consists of a complex of climate, soil, vegetation and animal life each of which, to a greater or lesser extent, affects and is affected by the others, both directly and indirectly. As we shall see, climate and soil influence vegetation and thereby the fauna. The soil results from the action of climatic factors and vegetation upon the rocks that make up the crust of the earth. Vegetation,to some extent, affects the climate and certainly ameliorates its adverse effects on the soil. Overgrazing, by destroying the vegetation, can thereby influence both soil and climate.

Climate

Climate may be interpreted in terms of the movement of air masses and the way in which these interact when they converge. After it has lain for some days over a relatively uniform surface of land or sea, the air itself becomes uniform with regard to temperature, humidity and stability. A 'front' occurs where two distinct air masses, with different characteristics, come into contact with one another. In temperate regions, fronts are usually characterised by differences in temperature between adjacent air masses but, in the tropics, there is usually no such contrast in temperature, the only differences being in stability and humidity. The distribution of air masses and fronts is closely related to the general pattern of the distribution of atmospheric pressure, and changes more with latitude than with longitude. This has been recognised since earliest times, so that names have been coined, such as 'doldrums' and 'trade winds', to describe the climate and different latitudes.

Winds

Solar radiation is the source of atmospheric energy, and its distribution over the earth not only controls climate, but is itself a climatic element of the greatest importance. The amount received at any latitude depends upon the angle at which the sunlight strikes the curved atmosphere of the earth, and the duration of day-length at that season of the year.

The annual total of solar radiation reaches a maximum at the equator and diminishes regularly towards minima at either pole. At the time of the equinoxes (about 21 March and 23 September each year), the sun's noon rays are vertical at the equator and the latitudinal distribution of solar radiation resembles that for the year as a whole. At the solstices (about 22 June and 22 December), however, when the noon rays of the sun are vertical on the 23½° parallels north or south, solar radiation is distributed very unequally in the two hemispheres. The summer hemisphere then receives two to three times the amount of solar radiation received by the winter hemisphere (Trewartha, 1968).

Temperature and precipitation are controlled by atmospheric pressure and winds. Each hemisphere has a maximum pressure throughout the year in the sub-tropics or 'horse latitudes'. Regions of lower pressure occur at the equator and the poles. The prevailing winds tend to blow towards them from the sub-tropical highs, along a gradient of pressure. These planetary winds are deflected by the earth's rotation, or Coriolis force, which acts at right angles to the direction of movement. Winds, like all other moving objects, no matter their direction, are deflected to the right of the gradient in the northern hemisphere and to the left in the southern hemisphere.

The equatorial trough or 'doldrums' is centred near 5° in January and around 12-15°N in July: it migrates through some 20° latitude between seasons. This migration brings seasonal rainfall and the formation of tropical storms. Since the annual mean equatorial trough lies near 5°N rather than on the geographical equator, this latitude is often known as the 'meteorological equator'. Although the distribution of atmospheric pressure on the globe is caused primarily by the decrease in temperature from the equator to the poles, and the effect of the earth's rotation, this general scheme (Fig. 4) is modified by the irregular shapes and distribution of the continents and oceans (Watts, 1955).

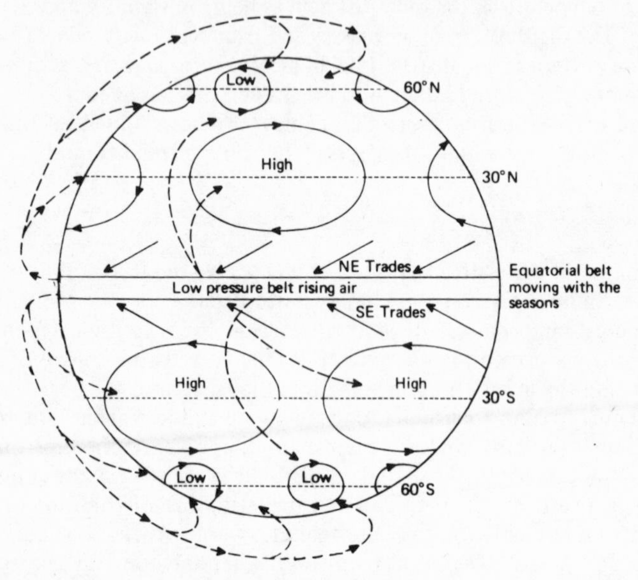

Fig. 4 Planetary wind systems. (Modified from Watts, 1955).

The main planetary winds are as follows:
(a) *The deep trades.* These are at their broadest in summer: during the winter they are restricted to a narrow belt along the equator. (b) *The equatorial westerlies.* These blow towards the equator in the hemisphere which is experiencing summer. In India they are responsible for the south-west monsoon. (c) *The sub-tropical easterlies.* These occupy the poleward margin of the tropics in summer, and cover most of the tropical belt in winter. (d) *The polar westerlies.* These blow towards the polar fronts which, with the doldrums and horse latitudes, form a third transition zone along a line of discontinuity. (e) *The polar easterlies.* These are shallow, fickle winds located in the high latitudes of each hemisphere.

The chief sources of uniform air masses are (a) The horse latitudes high pressure cells, especially over the oceans in summer but, to a lesser extent, over such large land masses as North Africa and Australia in winter. (b) The cold continental masses of Eurasia and North America in winter. (c) The polar regions. Air masses can thus be described as 'polar' or 'tropical', which characters determine their temperature conditions and capacity to hold moisture; and 'continental' or 'maritime'. These origins determine the extent to which their capacity to carry moisture is realised – that is, their relative humidity (Miller, 1965).

Circulation of the atmosphere

If the earth were not rotating, its atmosphere would circulate in the form of two gigantic convectional systems. Air warmed in equatorial and tropical regions would rise, flow towards the poles at high levels and there subside to ground level and give up its heat. At the same time, cold polar air would flow back towards the equator as surface currents. In tropical regions, where Coriolis effect* and rotational components are weak, atmospheric circulation does indeed resemble a simple convectional system. Nearly 60 per cent of evaporated moisture is furnished from the tropics and tropical cumuli have effects far beyond the regions from which they originate.

In middle latitudes, on the other hand, Coriolis force is strong and the westerly winds flow from a warm source towards a cold one. Heat transfer is accomplished less by steady flow than by strong, sporadic exchanges of air masses across the latitude belts. Airflow is variable at all heights and exchanges of air masses accompany a succession of cyclones and anticyclones which move from west to east. Circulation is sporadic nearer the poles, but its overall direction resembles that of the tropics more than that of the middle latitudes. In general, the movement of surface air is largely from colder to warmer regions. Consequently, polar and tropical air masses tend to clash within the middle latitude westerlies, with important effects on temperature, precipitation and storminess. In addition, since air tends to rise in middle and equatorial latitudes, and to settle in the sub-tropics and polar regions, clouds and precipitation are more frequent in the former (Trewartha, 1968).

Condensation and precipitation depend upon two variables: the amount of cooling and the relative humidity of the air. They form a continuous process, for strongly hygroscopic particles begin to attract water around them even at humidities as low as 75 per cent. The climates of the different zoogeographical regions of the world depend

---

* Winds moving in a straight line above the rotating earth follow courses curved relative to the land surface.

mainly upon radiation, temperature and precipitation.

Classification

The classical division of the earth, into torrid, temperate and frigid zones, was delimited by the mathematically and astronomically defined tropics and polar circles. It recognised the importance of temperature but ignored the effects of other factors such as the presence of land and sea. Innumerable combinations of climatic factors, acting on an infinitely variable topography, produce a bewildering number of geographical climates. Many classifications for them have been proposed, of which the more important are those of Supan (1896), and of Koppen (1931), first published in 1910.

Supan divided the world into 35 provinces, each characterised by a certain combination of climatic elements. He suggested that the mean annual isotherm of 20°C (68°F) should be regarded as the limit for hot climates and that the polar zones should be defined by the isotherm of 10°C (50°F) for the warmest month. This is a line of real significance for summer warmth is the factor of greatest biological significance in cold climates.

The same isotherms were adopted by Koppen in a more complete classification based on the duration of temperature in each region above, between or below these critical values as follows:

(a) *Tropical belt* – 12 months above 20°C
(b) *Sub-tropical belts* – 4-11 months above 20°C and 1-8 months between 10-20°C
(c) *Temperate belts* – 4-12 months between 10-20°C
(d) *Cold belts* – 1-4 months between 10-20°C and 8-11 months below 10°C
(e) *Polar belts* – 12 months below 10°C

This classification has the merit of simplicity, but it is arbitrary and leads to inconsistencies. Climatic boundaries are never sharply defined and there are transition zones, across which one type merges imperceptibly with its neighbours.

Koppen's scheme, although subsequently changed and modified, was accepted widely for more than four decades. Almost all schemes of climatic classification have subdivisions and boundaries based partly upon parameters of temperature and rainfall which are not significant in themselves. Indeed, 0°C (32°F) is unique because it marks the freezing point of water: no other temperature has any precise significance. The same is true of precipitation – there is no one particular value which is more important than any other (Trewartha, 1968).

Based upon annual and monthly means of temperature and rainfall, Koppen's (1931) classification accepts natural vegetation as the best expression of the totality of a climate. Consequently many of the climatic boundaries are selected with vegetation limits in mind. Koppen recognised that the effectiveness of rainfall on vegetation depends, not only upon the amount of precipitation, but also upon the intensity of evaporation and transpiration from the plants. He therefore tried to combine temperature and precipitation in a single formula indicating the effectiveness of rainfall. He did not use evaporation data *per se* but, nevertheless, introduced a factor which weighted precipitation unfavourably in areas where it fell mainly in summer. He did this because he regarded summer rainfall, especially in deserts, as being less effective than winter rainfall, owing to the higher prevailing temperatures at that season. (See discussion in Cloudsley-Thompson and Chadwick, 1964.)

The Koppen classification identified five main groups of climate, viz.: (a) Tropical rainy, (b) Dry, (c) Warm-temperate rainy, (d) Doreal, (e) Snow. Of these, only one, (b) Dry, is not thermally defined. Trewartha (1968) proposes five climatic groups based on criteria of temperature: (a) Tropical, (c) Sub-tropical, (d) Temperate, (e) Boreal, (f) Polar; and one based on precipitation, viz. (b) Dry (where potential evaporation equals or exceeds precipitation). A somewhat similar classification (Fig. 5) based on Miller (1965) and summarised in Appendix I has been adopted in the present volume.

More recently, Holdridge (1947) classified the vegetation of the world using temperature, precipitation and evapotranspiration (the sum of evaporation and transpiration) as climatic factors. In his scheme, biotemperature – the sum of mean monthly temperatures of those months with mean temperatures above 0°C, divided by 12 – is believed to be directly proportional to the potential amount of evapotranspiration that would take place if moisture were available. By plotting temperature, precipitation and potential evaporation, Holdridge defined boundaries for the major latitudinal and altitudinal zones of the earth and, within these, the major vegetation regions. Each polygon in his diagram (Fig. 6) corresponds either to a 'biome' or major plant association, or to a subdivision of a biome in other systems of classification. This system is in current use and clearly has great possibilities.

Altitude

Temperature decreases with altitude at a rate approximately 1 deg. C* for every 160 m of ascent. There are naturally considerable variations for local reasons. Moreover, the rate is usually less in winter than in summer, and less at night than during the day. High mountains experience regular temperatures below freezing and, subject to adequate precipitation, are perpetually snow-capped. Mountains also enjoy higher rainfall than lowlands in the same latitude (p. 150) and, in desert regions, form altitudinal oases (Cloudsley-Thompson, 1975b).

* In S 1 units, deg C = K.

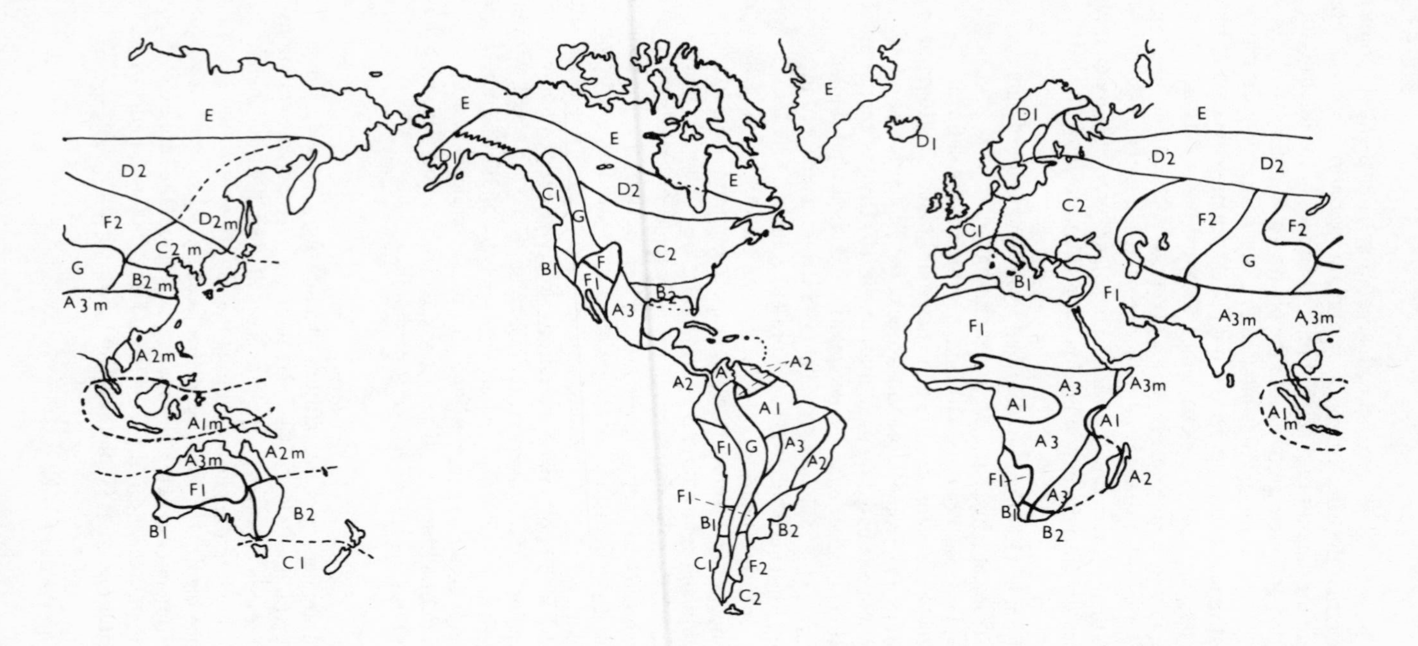

Fig. 5 Distribution of climatic types. (Lettering as in Appendix 1) (Modified from Miller, 1965). m = monsoon.

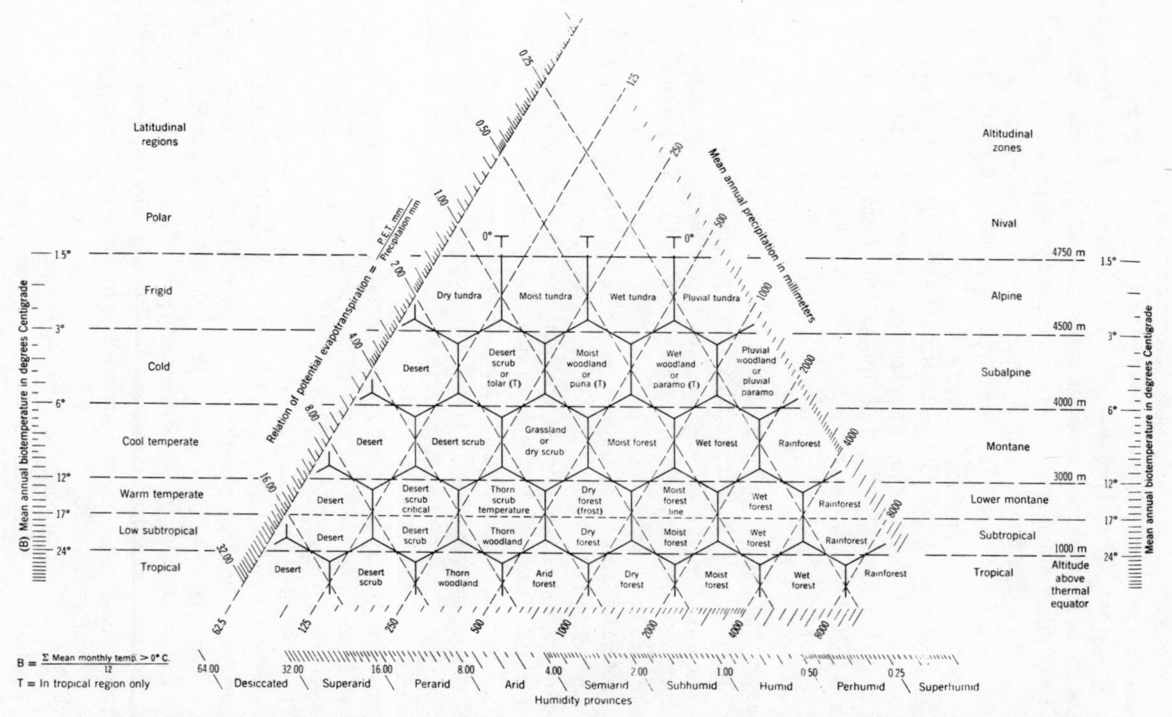

Fig. 6 Holdridge's (1947) system of classification of the world plant formations and life zones.

Although the temperature zones passed through in the ascent of an equatorial mountain are the same as those passed through on a journey from the tropics to the poles, conditions in the corresponding temperature zones are widely different. At higher altitudes there is a reduction in atmospheric pressure which has physiological effects on the fauna. Far more important, however, is the fact that the characteristic feature of insolation in the polar zone is the long summer day of oblique sunshine through a great thickness of atmosphere. The feature of insolation in alpine climates is the intensity of sunshine through a thin and very clear atmosphere (Miller, 1965).

The main biomes, to which subsequent chapters are devoted, are delineated more by their vegetation than directly by their climate. And vegetation is a product not only of climate, but of soil. At the same time, the soil is, to a large extent, produced by vegetation (Fig. 7).

Microclimate

It has long been known that organisms occupying the same general habitat may actually be living under very different physical conditions. The climate of the air near the ground is not the same as that measured in a meteorologist's Stephenson Screen at a height of 1.5 m above the soil surface; and the environment of a bird is very different from that of a centipede! Most of the animals inhabiting the earth lead hidden lives, as Savory (1971) points out. Some of them hide away from the sight of man for part of the time, emerging now and again when the coast is clear. Others are hidden from view throughout the entire cycle of their existence. The animals that we see around us – cows, horses, dogs, cats, eagles, crocodiles, sharks and hippopotamuses represent the familiar few. The teeming multitudes are to be found hidden in the soil, or beneath the debris that covers much of the earth's surface. The latter environment, the cryptosphere, provides a series of microhabitats that link their inhabitants with the environment in which they evolved many millions of years ago. (*See* p. 163).

The problem of controlling loss of water by evaporation is probably the greatest facing small terrestrial animals (Cloudsley-Thompson, 1962a). An important characteristic of the microclimates within the cryptosphere is that they tend to be more humid and stable than is the climate of the macrohabitat outside. The air between soil particles is usually saturated, even when the soil appears comparatively dry; while the temperature is warmer at night, and cooler during the day, than it is on the surface of the soil (Macfadyen, 1963).

The factors that influence microclimates are:

(a) *Topographical effects.* Shelter from wind results in the production of eddies and turbulence immediately behind the barrier that destroys normal laminar flow. Small particles such as snowflakes, leaves and seeds,

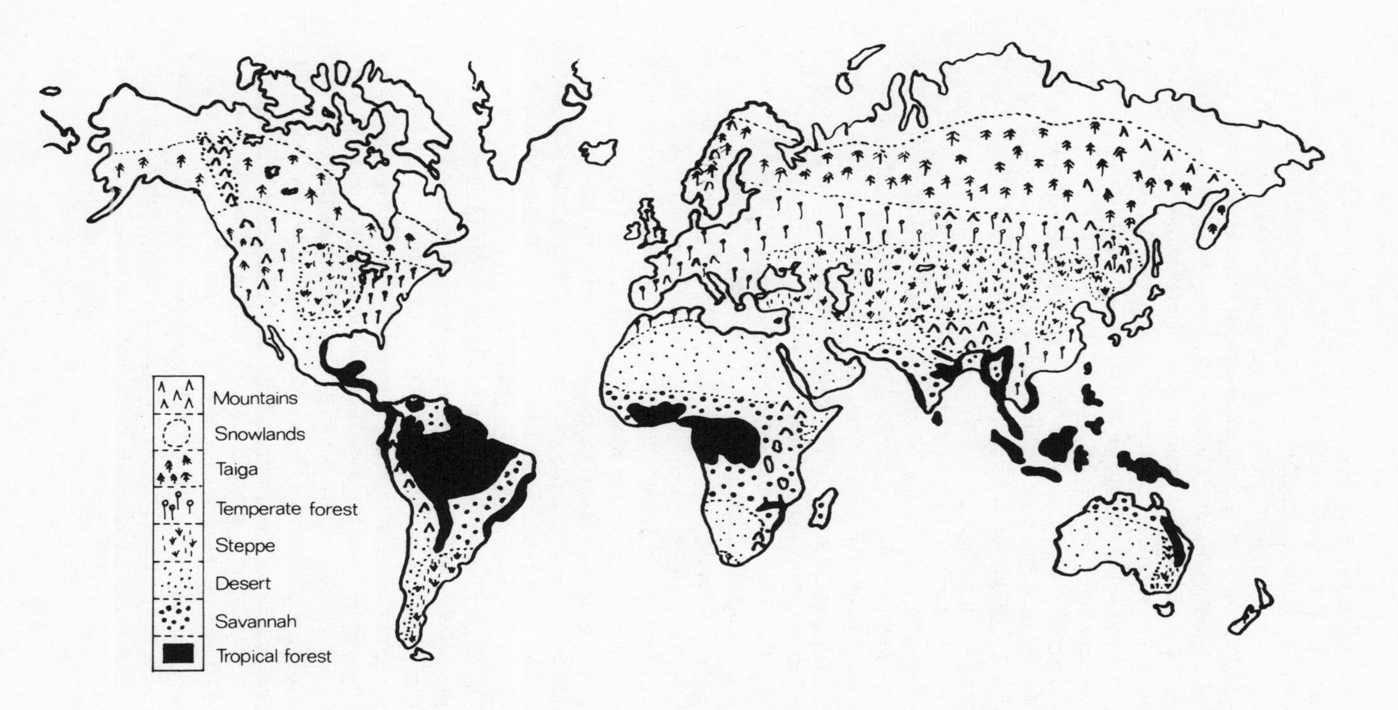

Fig. 7 Vegetational biomes of the world.

tend to be deposited in an area behind a barrier, such as a wall, for a distance extending ten to fifteen times the height of the barrier. In a detailed study of the microclimate, including both air and soil measurements in different directions and distances from a wind-break, Aslyng (1958) found that a slightly increased risk of late frost in the sheltered region was usually balanced by the higher humidity which reduced the actual damage to plants. No differences were detected in soil moisture, but potential evapotranspiration was reduced.

(b) *Temperature.* The climate of the air above the ground depends both upon the proximity and on the nature of the surface. As Geiger (1950) has emphasised, the level at which the highest day-time and lowest night-time temperatures are found may be either the surface of the ground itself, or the zone of maximum development of foliage. It is the surface at which absorption of the sun's radiant energy mostly takes place, and from which it is mainly radiated. The temperature on a sunny day at the surface of dunes in Finland reached a maximum of 47°C when the air 30 cm above was only 29°C and the temperature of the sand at a depth of 10 cm was 17°C (Krogerus, 1932). In measurements at Sukakangas in the south of Finland, soil surface temperatures of 50-60°C have often been recorded, the peak value being 63°C. In humus and litter, the rise in temperature is greater than on bare sand (Vaartaja, 1949). Sand surface temperatures often reach an extremely high level in the tropics. The world record is 84°C (183°F), registered on the Loango Coast near the equator and equalled on black Nile silt at Wadi Halfa, Sudan, in September. There is also an unsubstantiated record of 85°C (185°F) from Chinchosho in West Africa (Cloudsley-Thompson, 1965a).

The nature of the soil surface has a great influence on the amount of heat required to bring about a given change in temperature. Colour, to some extent, texture and water content, are of importance, the first two governing the ratio of absorbed to reflected energy; the last, as a result of the high latent heat of vaporisation and high specific heat of water, governing the rate of rise of temperature (Macfadyen, 1963).

(c) *Other factors.* Microclimate conditions are influenced not only by topographical effects and radiation, but also by humidity, wind speed and other physical conditions that affect the lives of plants and animals. Caves and rock fissures, for example, form the natural habitat of many cryptozoic animals because the microclimate conditions within them are relatively uniform. A wide range of microclimatic conditions may be available within a very short distance. By burrowing into the soil or entering a cave, desert animals can avoid the extreme heat and drought of the daytime. And, by leaving their burrows at dusk, they can also avoid peak temperatures below, for there is a considerable time-lag before heat begins to penetrate deeply into the soil.

Soil

Soils develop through weathering of the parent rocks that make up the crust of the earth. In addition to this mineral substrate, in which the vegetation takes root, soils include dead organic matter and humus, water and air. Weathering of the rocks to produce the 'parent material' of the soil is usually achieved by a combination of three processes:
(a) *Mechanical weathering,* in which breakdown of the rock itself occurs without chemical change;
(b) *Chemical weathering,* where an actual chemical change of the rock minerals takes place and new substances are formed;
(c) *Biological weathering,* where the agents of change are plants and animals.

Soil types
As the result of such weathering, a characteristic layered arrangement known as a 'soil profile' is developed. This depends largely upon the amount and kind of organic matter present and also upon the way in which water falling on the surface removes and redeposits the soluble constituents of the surface layers. Under humid conditions, soluble salts are 'leached' away but, where water is scarce, 'pedocals', or soils rich in calcium, are formed. 'Pedalfers' or leached soils, occur under humid conditions where rainfall is in excess of potential evaporation.

A soil profile consists of several horizons (Fig. 8) each having characteristic physical and chemical properties. At the surface there is

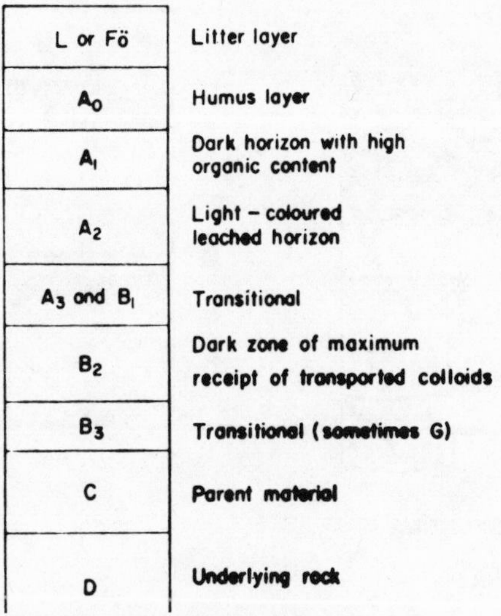

Fig. 8 Nomenclature of soil horizons.

frequently a layer of undecomposed material, the litter of L layer (often called Fö layer, from the Swedish term, Forna). Beneath this lies the humus or A_0 layer composed of amorphous organic matter which has lost its original structure. Then comes a varying number of A layers of true soil. The first of these, A_1, is a dark-coloured horizon containing a relatively high content of organic matter mixed with mineral fragments. It tends to be thick in savannah and thin in forest soils. The A_2 horizon is frequently ashy grey and is the zone of maximum leaching. The underlying B horizons tend to be darker in colour because they are enriched by iron compounds, clay and humus. A lighter coloured C horizon of parent material then grades into the D horizon of bed rock.

This description is applicable to 'podsols' (Fig. 9), acid soils with strongly acid hydrogen ion concentration (usually below pH 5.5) and excessive drainage. Podsols are developed on sandstones under conditions of moderately heavy rainfall. In 'brown earths' or 'brown forest earths', the profile is less uniformly coloured throughout, with a darker humus-rich A_1 horizon on top which grades into slightly lighter coloured subsoil. Brown earths are usually slightly acid and certainly never base-saturated. They are frequently developed over clays and drainage is often impeded to some extent.

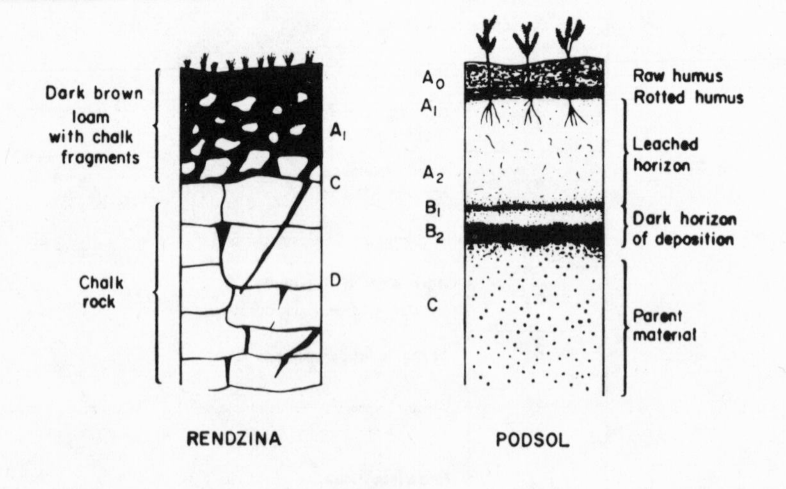

Fig. 9 Podsol and rendzina profiles.

Low pH results in an accumulation of litter in which bacteria are comparatively inactive and in which fungi, while they break down cellulose, have relatively little action on lignin. The 'raw humus' that results is sometimes called 'mor' and is often the product of coniferous trees.

Although the soils in tropical rain-forest do not become very acid, high temperatures throughout the year cause the clay materials to decompose quite rapidly, and the silica fraction especially becomes mobile. It is leached downwards, and either deposited in the material below, or else removed altogether with the drainage water. The upper horizons thus become relatively rich in sesquioxides and are mainly red in colour. This process, known as laterisation, is widespread in the humid tropics, although it does not occur everywhere. Some parent materials engender the formation of podsols, granite produces brown forest earths, and a great range of other soil types occurs in areas with impeded drainage.

On alkaline rock such as chalk and carboniferous limestone are developed 'rendzinas' (Fig. 9) or 'humus-carbonate' soils having no appreciable B horizon. Rendzinas are often base-saturated (having a pH of 7 or more) with an upper horizon that is usually dark brown, sometimes with a whitish tinge on chalk. This grades directly into lighter coloured parent rock. Rendzinas are usually shallow, freely drained and carry a typical vegetation.

The nature and quantity of the organic litter in soil depends largely upon the type of vegetation cover and its rate of decomposition. In alkaline or neutral soils, decomposition tends to be fairly rapid as a result of the activities of bacteria and fungi which here attack both lignin and cellulose. At the same time, larger soil animals such as earthworms and insects contribute towards the mixing of humus with mineral particles. The humus of such soils is called 'mull'', and is typical of brown earths and usually derived from broad-leaved trees. Intermediate between mull and the humus of greater acidity called 'mor', is 'moder', which has a richer and more varied fauna than mor, although plant remains are not broken down to the same extent as in mull. Moder tends to be considerably eaten into and mixed with faecal matter, yet is not matted together as raw humus and some mineral matter is also incorporated.

The pedocalic type of soil most similar to fully leached soils is called 'chernozem' or 'black earth', and is found under steppe and savannah grassland in central Europe and Asia, North America, Argentina, East and South Africa where, in spite of summer drought, the annual rainfall is over 25.5 cm (10 in.). The surface horizons appear dark due to the presence of organic matter which has become humified under conditions of high temperature and alkalinity. On the arid side of the chernozems a soil type known as 'chestnut earth' develops. The humus horizons are less well-developed than in chernozem soils, and the calcium carbonate deposits come near to the surface. Chestnut earths are covered with low grass steppe vegetation and occasionally some scrub.

In semi-arid regions where the rainfall varies between about 12.0 and 25.0 cm per year, the soil types usually formed are 'brown' or 'grey semi-desert soils'. The latter are sometimes called 'sierozems'. Sierozems may have even less than one per cent of organic matter in the surface horizon and calcium carbonate deposited on the soil surface. They often support low desert-scrub vegetation.

Desert soils, produced almost entirely by physical weathering, contain no humus and are little more than fragmented rock. The winnowing effect of the wind, sorting out particles of different sizes, transporting and depositing them elsewhere, results in the formation of three main types of desert: *hammada* or rocky desert, *reg* and *serir* or stony desert and *erg* or sandy desert. Sometimes the finest soil particles are removed so far by wind that they are deposited in the loess of the steppe-lands that border the desert.

Low-lying regions close to the equator where the annual rainfall exceeds 200-cm (80in.) are usually clothed in dense rain-forest whose determining conditions are a high, even temperature and abundant moisture. For the most luxurious development of rain-forest the precipitation must be distributed evenly throughout the year. In areas where the soil is periodically submerged, swamp jungle develops and this may in coastal regions take the form of mangrove swamps, for mangroves can flourish only where their roots are periodically submerged and uncovered by tidal action. In tropical and equatorial rain-forests there is little humus because high temperatures and humidity result in the rapid decay of organic matter. At the same time, lack of calcium owing to leaching possibly results in a poor invertebrate fauna, apart from the ever-abundant termites and ants which, by mixing humus with the soil, partially fulfil the functions of earthworms in temperate forests. In spite of the large quantities of vegetable matter consumed by termites, there appears to be little increase in soil fertility as a result. This is probably because the termites' digestion is very efficient and they eat not only their own excrement but also the corpses of dead termites. Thus little is left over to enrich the earth. Nevertheless, they move a great deal of soil and influence it considerably by their burrowings.

Soil profiles in part reflect features of surface relief as well as of rainfall and parental material. Shallow soils develop in hilly regions with accompanying excessive run-off and erosion. Flat land has little or no erosion so that a leached upper soil overlying a dense clay pan results.

Waterlogged horizons are usually greyish-green or greyish-blue indicating anaerobic conditions and the presence of organic matter.

Such soils are known as 'gley' and are often overlain by a horizon with rusty-brown mottling due to the presence of oxidized iron compounds. The layer affected is sometimes designated by the letter G or by (g) following the symbol for the appropriate horizon – e.g. B (g). Low-lying regions with poor drainage favour the accumulation of humus which may even form peat, containing more than 65 per cent by weight of organic material. In such anaerobic conditions, organic remains of plants become only partially decomposed (Cloudsley-Thompson, 1967a).

Mechanical composition and structure

Soils are often described in terms of the proportions of particles of different sizes that they contain. Particle size is not only an obvious characteristic, but it affects the growth of plants, their supplies of water and nitrogen, and the movement of animals. The composition of soil is determined by the proportions of the size fractions of particles of mineral matter, namely of sand (0.02 - 2.0mm diameter), silt (0.002 - 0.02mm diameter) and clay (below 0.002mm diameter). These particles are grouped and bonded into structural units which are themselves bonded together. The development of inter-particle bonds confers stability; the separation of structural units from each other determines their size and shape. Inter-particle bonds form most readily in the presence of clay, and the most stable bonds result from intergrowths of one kind or another. The separation of structural units may be caused by growing roots, ice formation, shrinkage during drying or the compression of air trapped when water seals the outer pores (Black, 1957). Russell (1957) and Wallwork (1970) discuss soil-forming processes, soil types and their production.

3 TROPICAL FOREST

A vast girdle of rain-forest encircles the earth between the tropics. It is bisected somewhat unequally by the equator, so that rather more of its area lies in the northern than in the southern hemisphere. Not only is the forest interrupted in many places by mountain ranges and plateaux but, in the more densely populated regions of the tropics, it has been replaced so completely by cultivation that it is impossible to reconstruct the original climax vegetation (Richards, 1952).

The largest continuous mass of rain-forest is found in tropical America, with the huge Amazon basin as its centre. This extends west to the lower slopes of the Andes and east to French Guiana, south into the region of the Gran Chaco and north, through the eastern side of Central America, into southern Mexico.

The central African evergreen rain-forest extends from the Congo basin throughout equatorial west and central Africa with many breaks of savannah. Rain-forest is found on the Zambesi and is the natural climax in part at least of many of the islands of the Indian Ocean including the east coast of Madagascar; but very little primitive forest now remains in these places.

In the eastern tropics, rain-forest extends from Ceylon and western India to Thailand, Burma, Malaysia, Indonesia, New Guinea, etc., and the north coast of Australia. This is the tropical forest formation richest in plant species (Fig. 6). Tropical rain-forest, especially in equatorial regions, constitutes the most luxuriant of all vegetation types.

Climate

Shifting cultivation, by which over 200 million people obtain their livelihood, has already destroyed much of the world's primary rain-forest and, in many cases, has changed the entire ecosystem. Except for the tallest, which are usually spared, the trees are felled, piled together, and burned. A clearing is thus formed, littered with charred logs and stumps about 2m high. Crops, such as cassava, are then grown in the soil enriched by ashes from the fire. After a year or so, the ground becomes leached, loses the little fertility it had, and a new area has to be cleared.

A deep humus layer is seldom to be found in tropical forest. When exposed to the elements, the shallow topsoil is soon eroded. This, in turn, reduces the capacity of the ground to retain moisture. The secondary forest which springs up in deserted clearings has a different floral composition than that of primary forest. The number of tree species is reduced from several hundred to twenty or thirty: these trees tend to be fast growing species with deep roots that can reach down to the lowered water table into which have been leached the minerals necessary for growth. After fifteen to twenty years, shifting cultivation claims the young secondary forest, and the whole process is repeated. After many such repetitions, the forest becomes permanently degraded and an aritifical savannah landscape is produced. This has happened over much of East Africa, where agriculture has been supplanted by pastoralism. Overgrazing and compaction of the soil by the trampling of domestic stock can then cause erosion and even create near-desert conditions in regions with high rainfall.

Forest crops include oil palm, coffee, cacao, vanilla, nutmeg, sweet potatoes, cassava, yams, rubber and so on. Although plants grow rapidly in the tropics, even in such a wet climate as that of the Philippines, only 15 per cent of farm land is cropped twice in a year and in India, with the largest irrigated area of any country in the world, the figure is only 13 per cent (Masefield, 1970).

Conditions necessary for the production of tropical rain-forest are heat and moisture. The mean temperature needs to be around 26.5°C (80°F), rarely sinking below 21°C (70°F) or exceeding 32°C (90°F). The moisture must be due to rain, with an annual minimum exceeding 150cm (60in.), more effectively 200cm (80in.) and distributed rather evenly throughout the year. The dry period, if there is one, should not amount to more than two or three months, lest there be a standstill of the vegetation, causing deciduous leaves to fall and other changes in the general aspect of the forest to take place (Gadow, 1913).

A distinctive feature of tropical humid climates is the fact that the average daily range of temperature exceeds by several times the difference between the warmest and coolest months of the year. The rays of the sun at noon are never far from the vertical, and days and nights vary little in length from month to month. Monthly average temperatures are not excessively high, but the climate is characterised by its uniformity and monotony. The abundance of cloud and heavy forest cover serve to prevent excessively high temperatures such as occur in tropical deserts.

The mean temperature of the hottest month at Belem in the Amazon basin is only 26.5°C (80°F) and, at most stations on the Congo, it is slightly less. The minor differences between the warmest and coolest months are determined, not so much by the position of the sun, as by the amount of cloud. The diurnal range is several times greater than the annual range of temperature in the wet tropics and equatorial regions of the world. At Bolobo on the Congo, for example, the average daily range is 8 deg. C* whereas the annual range is only 1 deg. C. Even the daily extremes of temperature are not great. The average of the daily maxima at Belem is only 33°C (91°F), that of the daily minima is 20°C (68°F) (Trewartha, 1968).

As already mentioned, precipitation is heavy in the wet tropics, and a large proportion of cloud is of the cumulus type. In equatorial lowlands, where rainfall is heaviest, the drenching rain may reach an average of 400 cm (160 in.) per year. The world record for lowland precipitation is an annual average of 1017 cm (403 in.) in Debundja, at the foot of Cameroons peak. The force of the downpour is an important ecological factor, for thunderstorms of great violence are frequent and the rain descends with a suddenness and volume unknown outside the tropics. To this may be related the fact that many rain-forest birds nest in the holes of trees. Most animals, in fact, shelter from the bombardment of the rain. Further away from the tropical belt, there are usually one or two wet periods during the year, alternating with dry spells. Where these increase in length, evergreen rain-forest is replaced by deciduous savannah with a different flora and fauna.

Vegetation

Although tropical and equatorial rain-forest varies greatly in different places according to altitude, rainfall, soil and insolation, there is an astonishing similarity throughout this vast region with regard to general growth form, the luxuriance of the vegetation and the great multiplicity of species among the trees present. For example, in temperate mixed forests, ten to fifteen species of tree may be present. In contrast, 400-500 tree species and some 800 species of woody plants have been described from the forest of the Cameroons. Competition is so severe in the optimum conditions of warmth and moisture that a local clearing, as may be caused by the destruction of a single large tree, can provide temporary accommodation for numerous small species that are otherwise rare or impoverished.

Of the many thousands of trees present, nearly all are evergreen, casting their leaves and growing new ones continuously, so that they are never leafless. The few species that shed all their leaves together do so at irregular intervals. The majority of the trees, though extremely diverse taxonomically, have dark green, leathery leaves similar to those of laurel.

The main components of the tropical rain-forest are as follows:
(a) *The forest trees* which form the 'roof' or canopy. As they push up towards the sunlight, they form a series of indistinct layers, each a miniature life zone in which animals live and feed. There are usually

* In S1 units, deg. C = K.

from three to five such layers in mature tropical forest. The highest consists of scattered trees which tower above a closed canopy layer formed by the crowns of tall trees. Many of them have buttresses, whose function may be to provide improved support without increasing expenditure of energy on wood (Smith, 1972). Below this canopy is the third or middle layer, formed by smaller trees whose crowns do not meet. The fourth layer is composed of woody and herbaceous shrubs. Finally, there is a ground layer of non-woody herbs and tree seedlings. Very little light reaches this layer. The rain-forest canopy averages 36.5m (120ft.) in height, the shrubs 2-4.5m (6-15 ft.), and the herb layer about 0.5m (2 ft.) (Fig. 10).

(b) *Herbs,* etc. occur where the tree strata are not too dense. Their stems tend to be sappy and brittle, no doubt because they have little lignified tissue and, in some species, the stem is supported at the base by aerial 'prop' roots. The herbs of the rain-forest are varied in their foliage, and contrast with the monotonous uniformity of the trees and shrubs; their leaves are usually thin and soft in texture. In a few species the leaves have a velvety surface. More often they have a variegated pattern of white or pale green. Flowers are usually inconspicuous.

(c) *Climbers,* vines or lianas, thin as ropes or as thick as a man's thigh, vanish like cables into the mass of foliage overhead. They are especially abundant on river banks and where the tree canopy is thinner. Although climbers comprise a fairly well-defined plant form, weak-stemmed trees and half-climbing shrubs are transitional between them and fully independent plants. Similarly, no sharp line can be drawn between climbers and epiphytes, many of which have climbing stems or began their life rooted in the ground and only later lose their connection with it.

Climbing plants are usually classified, according to their means of attachment, into 'scramblers', which have no specialised organs; 'twiners', whose tips perform revolving movements in a constant direction; 'root-climbers', which attach themselves with specially modified aerial roots, and 'tendril-climbers'. These are the most specialised and possess organs that are sensitive to the presence of a support and fix themselves actively to it. From an ecological point of view, the method of climbing is less important than the size of the climber and the maximum height that it reaches. Rain-forest climbers fall into two groups: large, woody lianas which reach the crowns of trees in the top stories, and are therefore exposed to the sunlight when adult; and much smaller, mainly herbaceous forms, which seldom emerge from the shade of the undergrowth. These two groups differ not only in their microclimatic environment, but in their relations to other plants.

(d) *Stranglers* form one of the most remarkable groups of rain-forest plants and have no parallel in Paeaearctic forests. They begin life as

Fig. 10 Rain-forest stratifications showing five layers of vegetation.

epiphytes but later send down roots to the soil. Eventually, they become independent, often killing the trees that originally supported them. 'Strangling-figs' (*Ficus* spp.) are abundant in the Ethiopian, Indo-Malaysian and Australian rain-forests. In South America, the most important genus is *Clusia* which, likewise, is represented by numerous species.

(e) *Epiphytes* grow on the trunks, branches, and even on the leaves of trees, shrubs and lianas. In a closed forest, the epiphytic habitat is the only 'niche' available for plants that combine small size with relatively high demands for light. They suffer from a precarious water supply and lack of soil, for they depend on their 'host' plants only for mechanical support: only Loranthaceae (mistletoes) are parasitic as well as epiphytic. Epiphytes play an important part in the forest ecosystem as microenvironments for small animals. The spaces between their leaves are frequently inhabited by a vast array of insect larvae, planarians, earthworms, snails, woodlice, centipedes, millipedes, termites, grasshoppers, earwigs, ants, scorpions, spiders, tree-frogs, lizards, snakes and so on. Many epiphytes are so constructed that they collect in their roots a substitute soil derived from the dead remains of other plants and often assembled by ants.

(f) *Saprophytes* include bacteria, fungi, orchids, gentians and so on. These obtain their nutriment from decomposing organic matter on the forest floor or between the buttresses of trees where dead leaves tend to accumulate. Excepting some of the orchids, saprophytes tend to be inconspicuous and easily overlooked.

(g) *Parasites* are either terrestrial root-parasites or arboreal epiphytes. Only two families of root-parasites, the Balanophoraceae and the Rafflesiaceae are represented in tropical rain-forest. The latter include the Malayan *Rafflesia* spp., famous for their gigantic flowers: the former, a number of genera such as *Helosis* and *Thonningia.* Epiphytic semi-parasites consist of a single family already mentioned, the Loranthaceae, which has a very wide range of hosts. Some species are hyperparasites, occasionally even to the second degree – a parasite on a parasite on a parasite (Richards, 1952).

Monsoon forests

Where marked dry seasons occur in otherwise humid tropics the vegetation presents a more varied appearance. Monsoon and similar seasonal climates tend to engender forests dominated by a wide range of deciduous trees. These are found on the margins of the equatorial rain-forest. Their vegetation is more open and less luxurious; the canopy tends to be much interrupted and the undergrowth is denser. Such forests grade imperceptibly into savannah-woodlands, which are discussed in the following chapter.

Monsoon forests are found in India, Burma, Indo-China and southwards to northern Australia as well as on the margins of tropical rain-forest in Africa, Madagascar, Indonesia, and central and South America. Their vegetation is not so luxuriant as that of tropical rain-forest, and tends to be more open. Many of the trees shed their leaves in the dry season, but this is often the period of their flowering. Consequently, the forest does not appear entirely lifeless even at that time of year.

The trees tend to have thick bark, exhibit growth rings and lack buttresses. Their trunks are massive but fairly short, seldom exceeding 35m (115 ft.) in height. Climbers are fewer and smaller than in tropical rain-forest, and vascular epiphytes are normally found only in the canopy. Consequently, the undergrowth is often luxuriant, consisting of shrubby thickets or tall grasses (Polunin, 1960).

Swamp forests

Much of the South American rain-forest is inundated intermittently. At certain seasons, many hundreds of square km of the Amazon and Orinoco basin are submerged and the soil is wet and marshy. These regions are not so richly furnished with lianas and climbers as is drier forest. Aloft, they are scarcely missed, however, for their place is taken by a host of epiphytic and parasitic plants which crowd the branches. The comparative bareness of the trunks of the great trees, like the unadorned pillars of a vast cathedral, adds only to the impressiveness of the scene (Haviland, 1926).

Mangrove swamp forests

Mangrove is characteristic of muddy tropical sea-shore. The term is used to describe a group of woody plants inhabiting tidal lands in the tropics. Mangroves form a good example of convergent evolution because, although they have a number of marked structural characteristics in common, they belong to widely different families of plants. They all have thick, leathery, evergreen leaves of simple shape. Some have prop-roots high up on the stem and often breathing roots also, with minute lenticels through which air diffuses, sticking up vertically from the surface of the mud. They also have a tendency for their seeds to germinate while still attached to the parent plants. The radicles grow straight downwards so that they are driven into the mud when the seeds fall. West African mangroves are of the same Atlantic species as occur on the tropical coasts of eastern South America and include the white mangrove *(Avicenna germinans)*, and the red *(Rhizophora mangle)*. The species found on the eastern seaboard of Africa and Madagascar, on the other hand, have Asian affinities. They include *A. marina* and *A. mucronata.* Nevertheless all mangrove swamps look alike and take the form of dense, evergreen forests of medium-sized trees with bright green, glossy leaves (Cloudsley-Thompson, 1969).

Mangroves often extend in brackish swamps and lagoons for some distance inland, forming a fairly continuous fringe, or occupying islets between which run sluggish tidal streams, for brackish or occasional salt water is necessary for their development. Compared with the rain-forest behind it, the mangrove swamp is a light and airy place. The trade winds from the sea blow through it and the sunshine can enter.

Fauna

The inhabitants of the tropical rain-forest can be divided into a number of ecological groups, according to their ways of life. For instance, some of the mammals have acquired arboreal habits and are adapted for climbing trees. Others are terrestrial and have to be able to push through dense undergrowth. Subterranean forms are relatively scarce compared with the numerous burrowers of savannah and steppe. Cursorial birds are naturally less common than in open country, but arboreal species are well represented. Many of the reptiles and amphibians have become adapted for climbing (Chapter 13).

The cryptozoic fauna of the forest floor is rich in species and forms a distinct community of animals, which is discussed in Chapter 11. Flying insects may not exhibit any marked adaptations to the environment, but their richness in species reflects the luxuriance of the vegetation that supports them: bumble-bees are perennial in the agsence of a seasonal climate. Nevertheless, although the number of animal and plant species found in the humid tropics is higher than in any other region, the numbers of individuals, except, perhaps, of social insects, tend to be low and to remain fairly constant. Predator-prey oscillations, such a feature of ecologically simple environments like the Arctic tundra, are buffered by the extreme ecological complexity of the rain-forest which has the greatest natural inherent stability of any region in the world (Cloudsley-Thompson, 1969).

The diversity of species present in rain-forest may be one of the factors limiting the size and abundance of larger animals because few single, constant sources of food are available, except ants and termites. All the larger mammals, even those that are arboreal, tend to range rather widely. This is illustrated by chimpanzees *(Pan troglodytes).* As Reynolds (1965) found in the Budongo forest of Uganda, these animals move from one part of the forest to another as the fruits of different trees ripen. This requires a flexible organisation to enable them at one time to be scattered all over the forest, at another to be concentrated at one spot. In contrast, lowland gorillas *(Gorilla gorilla)* can find almost anywhere the pithy stems and roots on which they live; so they travel in compact groups, moving slowly and quietly, and feeding as they go.

Orang-utans *(Ponto satyrus)* in Sumatra move only a few hundred metres per day, feeding on fruit and vegetable matter as they go. They inhabit equatorial forest that does not experience seasonal drought as do the forests of Uganda. The males are solitary and the females usually live with a single sub-adult or infant. The suggestion has recently been made that polyspecific associations of rain-forest primates, by increasing the effectiveness of group size, give advantages in food location and avoidance of predators without increasing interspecific competition for food and competition between males for females (Gartlan and Struhsaker, 1972).

Seasonal rhythms

In regions of evergreen forest where the rainfall is distributed evenly throughout the year, there is little seasonal change of climate. Annually recurrent floral cycles are displayed by only ten of 45 species of forest trees in Malaya, although the community as a whole shows some regular seasonality (Medway, 1972). Breeding is therefore not confined to a single season of the year and many birds, perhaps a majority, raise two clutches every twelve months. Nevertheless, although they may not be immediately apparent, seasonal rhythms of reproductive activity are to be found in most species. For example, birds can be found breeding in all months of the year in Uganda although, for any particular species, breeding may be restricted to a comparatively short period or periods (Owen, 1966).

In the rain-forest regions of tropical Africa, birds are the only major group of animals for which there is a reasonably complete picture of breeding seasons. Away from the influence of the great lakes, there are two relatively well-defined rainy seasons in East Africa. Birds tend to breed twice annually, either in the dry seasons or during the rains, depending on the species and on the seasonal availability of food for the young. In the more humid climate of the Congo, where rainfall is well distributed and temperature fluctuations are small, many species of birds seem to be non-seasonal. Nevertheless, even here, the breeding activity of passerine birds tends to be less during the drier months of November – January, but there are many exceptions. Woodpeckers breed especially at this time of year, while doves and birds of prey seem to avoid the wetter months of February-May (Chapin, 1932). For a recent popular account of the problems of seasonal breeding in the unvarying climate of rain-forest, see Smith (1970).

In temperate regions, the reproductive cycles of birds are controlled by seasonal changes in the length of daylight. These act as 'proximate' factors which function as signals heralding the approach of a suitable season. The ultimate determination of the reproductive cycle, however, depends upon the survival of progeny to an age at which they can reproduce. Natural selection favours the gene-complexes of those individuals that produce their young at the most propitious season (Baker, 1938). Photoperiodic responses are modified by a host of environmental inhibitors and accelerators, such as temperature, weather, food supply and interactions of behaviour. Variations in the photoperiod do not provide a reliable proximate stimulation in tropical and equatorial regions, and many species of bird that breed there have evolved a response to other environmental releasers, such as rainfall or the presence of green vegetation.

The continuous abundance of food, and relative absence of inhibitors such as drought, often permits the abandonment of a more or less precisely timed annual cycle in equatorial regions. Nevertheless, experiment has shown that birds still retain the capacity to respond to photo-stimulation in regions where environmental conditions remain constant and continuous breeding is feasible. In the absence of any proximate stimulus, the bird displays an autonomous cycle. But external pressures, such as predation, may lead to a high degree of synchrony in breeding, initiated by involved social displays, as in the sooty tern *(Sterna fuscata)* of Ascencion Island whose breeding occurs every 9.6 months. The physiological adaptations regulating the breeding cycles of birds, and their ecological significance, has been reviewed in considerable detail by Lofts and Murton (1968), to whose excellent article the reader is referred for further information.

The breeding seasons of mammals are less well documented than are those of birds. Although many species appear to breed through the year (Harrison, 1955), however, distinct peaks in breeding activity have often been found when large samples have been examined (Owen, 1966). Not only is breeding strongly seasonal in the fruit-bat *(Eidolon helvum)* at Kampala, but there is also delayed implantation so that the young are born in February and March (Mutere, 1965). Baker and Baker (1936) likewise found that, in spite of the almost unvarying temperature of the New Hebrides Islands (about lat. 15°S.) *Pteropus geddieri* copulates only during February and March, and the females, with few exceptions, give birth to their single young in August or September. As Allen (1939) pointed out, breeding begins at the time of year corresponding to the northern spring. This fact gives a clue to the process of adaptation to the winter season of bats living in more northern regions. On the whole, these agree in an autumnal copulation, but the sperms are stored during the period of hibernation and the ova are not fertilised until the spring.

In contrast, the long-fingered insectivorous bat *(Miniopterus australis)* copulates in that part of the year corresponding with the southern spring. Fertilisation and development of the embryo proceed without delay and there is no evidence of prolonged storage of sperms by the female (Baker and Bird, 1936). This species is a close relative of the European *M. schreibersii* which is exceptional in that both copulation and ovulation take place in autumn while the embryo develops very slowly during the winter.

Elsewhere I have discussed possible timing mechanisms for seasonal rhythms of reproduction in equatorial regions where there is no regular change in the period of dusk and dawn, and suggested that a lunar cycle might impart great accuracy to the entrainment of an internal rhythm with an approximately yearly periodicity (Cloudsley-Thompson, 1961a).

A marked influence of seasonal precipitation on the breeding of tropical mammals has been demonstrated in a number of instances. For example, many small mammals in Uganda have been found to show little reproductive activity between July and September, although some species appear to breed throughout the year (Delany, 1964). Peaks of breeding have been demonstrated in forest rodents such as *Praomys morio* and *Lophuromys flavopunctatus* at the end of the rains and the beginning of the dry season (Southern and Hook, 1963b). The multimammate rat *(Rattus natalensis)* breeds mainly at the end of the rainy season in East Africa as in Sierra Leone, the Congo, and elsewhere (Chapman, Chapman and Robertson, 1959).

Recent work in the Congo equatorial forest has shown that the squirrel *Funisciurus anerythrus* breeds at the peak of the early rains whereas pregnant females of *Paraxerus boehmi* are distributed evenly throughout the year. Whereas *F. anerythrus* is predominantly ground-dwelling, *P. boehmi* occurs mainly in bushes and small trees, and has a more insectivorous diet. In Muridae *(Praomys jacksoni, Lophuromys flavopunctatus* and *Hybomys univittatus),* the combined figures for the monthly incidence of pregnancy is highest during the wetter months of the year in forest. In abandoned cultivations, however, there is a more even spread of reproduction, with no clear seasonal pattern. In such places the occurrence of cultivated plants provides a constant supply of food throughout the year (Rahm, 1970). These findings implicate diet as a factor affecting reproduction, but whether it is proximate or ultimate has not been established.

The dikdik *(Rhynchotragus kirkii)* shows two peaks of intensive breeding in Tanzania, one at the beginning of the rains in November and one at the end of April. The gestation period of these antelopes lasts about six months, however, and most females become pregnant twice a year (Kellas, 1955). Elephants show seasonal peaks in breeding (Perry, 1953), as well as red-tailed monkeys *(Cercopithecus ascanius),* but not other *Cercopithecus* spp., nor species of *Colobus* and *Papio* according to Haddow (1952). The hippo *(Hippopotamus amphibius)* is seasonally polyoestrus in Uganda, with peak numbers of conceptions in February and August. The gestation period is about 240 days, so that the majority of young are born during months of high rainfall (Laws and Clough 1965).

On the other hand, the grysbok *(Raphicerus sharpei)* in Rhodesia shows no marked seasonal peak in breeding, and this must be true of many other mammals in tropical Africa (Cloudsley-Thompson, 1969). Thus, the angwantibo *(Arctocebus calabarensis)* has been recorded as breeding throughout most of the year in Gabon, while the related *Galago senegalensis* of the savannah reproduces only at the beginning of the rains.

Examples of seasonal breeding in equatorial regions of reptiles, amphibians, fishes and invertebrates are cited by Cloudsley-Thompson (1961a, 1969), Owen (1966) and others. Enough has been said, however, with reference to the better documented birds and mammals, to show that the comparative absence of distinct seasons in the wet tropics has not necessarily resulted in the loss of seasonal reproductive behaviour in animals. Indeed, the photoperiodically controlled seasonal reproduction of temperate bird species may well have evolved from the autonomous cycles of tropical species (p. 124).

Diurnal rhythms

Although the difference between day and night is less marked in the gloomy confines of the rain-forest than it is in open country, striking contrasts in the fauna still occur. In the South American forest, for example, curassows (Cracidae), tinamous (Tinamidae) and humming birds (Trochilidae) are active during the day. Butterflies flutter through the undergrowth and anolis lizards dart up the tree trunks. Larger animals such as brocket deer *(Mazama* spp.*)*, coatimundis *(Nasua* spp.*)* and agoutis *(Dasyprocta* spp.*)* are also about and, occasionally, howler monkeys *(Alouatta* spp.*)* venture low enough in the forest canopy to be seen from the ground.

At night a new array of animals takes over. Silky anteaters *(Cyclopes didactylus)*, woolly opossums *(Philander* spp.*)*, kinkajous *(Potos flavus)* and armadillos emerge from hiding in search of food. Bats fly silently through the trees, jaguars *(Panthera onca)* and other large predators set out on their nocturnal hunts, while in the clearings screech owls (Strigidae) and nightjars (Caprimulgidae) swoop near to ground level (Richards, 1970).

Diurnal rhythms of activity are found in nearly all terrestrial habitats, for all living organisms appear to possess a 'biological clock' or time sense, and most animals maintain a fairly inflexible routine of activity and rest (Cloudsley-Thompson, 1961a).

In the cathedral-like stillness of the African rain-forest, the chirping of insects and echoless calls of birds such as emerald cuckoos *(Chrysococcyx cupreux)*, bee-eaters (Meropidae) and grey parrots *(Psittacus erithacus)* sound high among the giant trees, while the queer call of the white-tailed hornbill *(Bycanistes sharpei)* resembles a mixture between a whistle and the bark of a dog. At dawn and dusk, however, the uncanny silence of the daytime is broken, and the forest resounds with innumerable strange cries. On account of limited visibility, intercommunication is perforce largely vocal (Cloudsley-Thompson, 1969).

For the zoologist it is more rewarding to explore the forest at night with the aid of an electric head-torch than to do so during the day. Reflection from the tapetum of their eyes will reveal amphibians, reptiles and small mammals that are never to be seen in daylight. We turn now to another peculiarity of the fauna of rain-forest, viz. a tendency for amphibians to reproduce away from water.

Terrestrial breeding in Amphibia

Many of the amphibians in rain-forest carry their eggs and tadpoles around with them. For instance, the Surinam toad *(Pipa pipa)* of neotropical forests is almost entirely aquatic, but does not spawn into the water. The eggs are glued to the back of the female where they sink into invaginations of the skin. As development proceeds, each egg becomes concealed in a pouch or cavity with a lid. Still other frogs carry their eggs about until the young emerge as adults (Fig.11) (Hesse, Allee and Schmidt, 1951).

Arboreal frogs of the genera *Rhacophorus* in the Old World and *Phyllomedusa* in the New, lay their eggs on land. In the South American *Phyllomedusa hypochondrialis,* the breeding pair bend or glue together the margins of a leaf to form a funnel into which are passed the eggs and sperm. Other species place their eggs in leafy sacs on branches that overhang water, into which the larvae fall on hatching. *Afrixalus dorsalis* of West Africa lays its eggs on vegetation above water but glues them together with slime in which they develop, while *Leptolepis anbryi* deposits its eggs on the ground where the larvae develop in damp soil. Many frogs lay their eggs in miniature ponds formed by the leaves of bromelias and other epiphytes. Here the tadpoles live until they metamorphose.

The female *Rhacophorus reticulatus* of Ceylon attaches the eggs, about twenty in number, to the undersurface of her belly. *Rh. maculatus* of S.E. Asia and *Rh. schlegeli* of Japan lay their eggs in a foamy mass of oviductal secretion in holes on the margins of pools above the surface of the water. At first, the froth is elastic and sticky, but it gradually sinks down, becomes liquid and ultimately runs out of the hole (Gadow, 1901). These tree-frogs beat their egg masses into a foam with the hind legs. The outer surface then dries, forming a crust inside which the tadpoles develop. In two species which increase the yolk content of their eggs, these are no longer beaten into a foam. The habit of making a foam nest has evolved independently in Bufonidae, but here the eggs

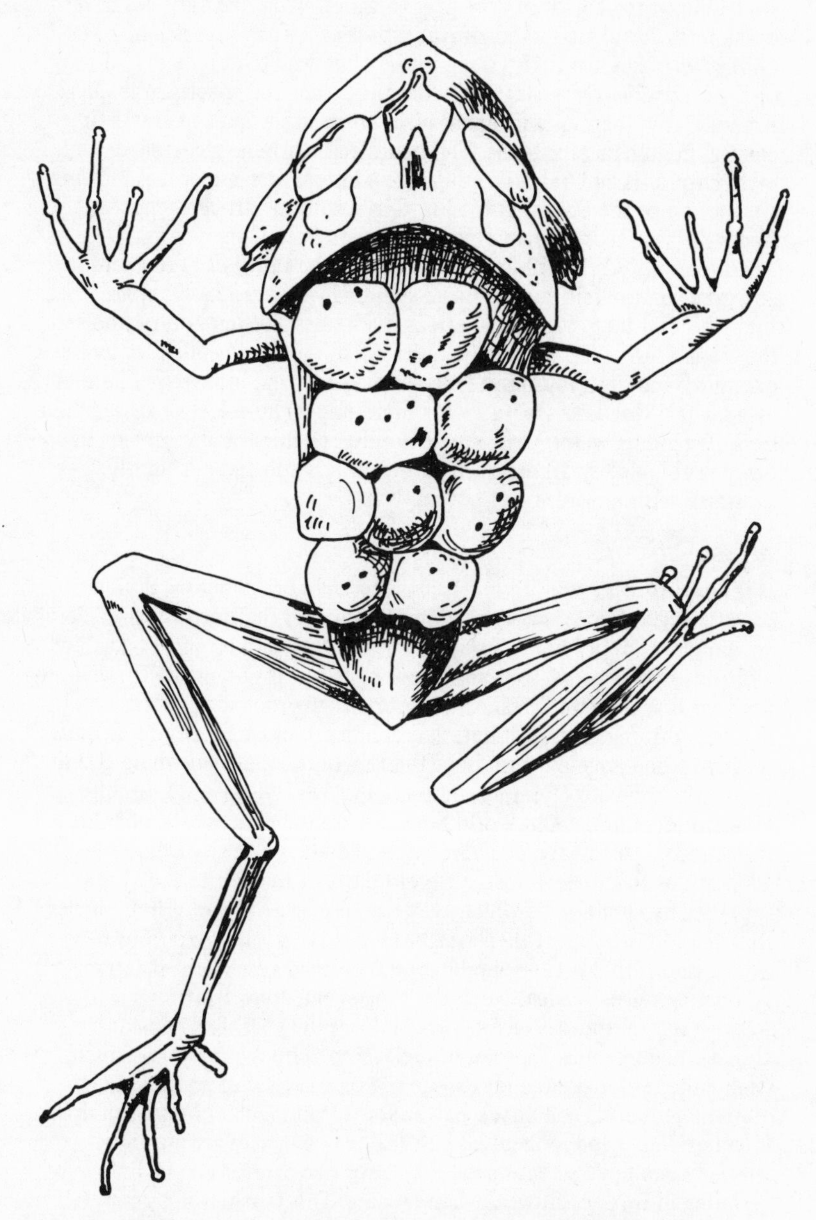

Fig. 11 Female of the tropical American frog. *Hemiphractus bubalus* carrying eggs attached to her back (After Hesse *et al.*, 1951).

are laid in contact with or very near to water, while the tadpoles never develop the larval characters of *Rhacophorus.* African species of *Chiromantis* care for their young in the same way (Noble, 1931). Frogs of the genus *Eleutherodactylus,* with more than 100 species in tropical America, glue their eggs to a leaf, place them in the axils of leaves or conceal them beneath stones. The young pass through a curtailed metamorphosis and hatch as miniature adults. The same occurs in *Rana phrynoides* of the Solomon Islands. Species of the African genus *Nectrophrynoides* are viviparous and give birth to living young.

Haviland (1926) attributes such traits to the rarity of permanent lakes and ponds in the rain-forest and of standing water generally, apart from the rivers and their permanent streams. Puddles and temporary pools in the tropics, however, speedily acquire a rich fauna of beetles, water-bugs, dragonfly and dipterous larvae (Rzoska, 1961), and it may well be that avoidance of predators is, in fact, a more important selective factor than lack of standing water. Certainly it is only possible for amphibians to become completely arboreal where the air is warm and constantly saturated with moisture.

Arboreal life

It is not surprising that arboreal species should be numerous in rain-forest, because vegetation is scanty at ground level, and the tree-tops provide an abundant supply of such staple foods as fruit and termites (Haviland, 1926). It is because seasons are absent and fruit is obtainable throughout the year that the rain-forest can support frugivorous forms such as parrots, fruit-pigeons, fruit-bats and chimpanzees which would be quite unable to find a regular supply of their favourite diets anywhere else in the world. Toucans (Ramphastidae) in the New World and hornbills (Bucerotidae) in the Old World both exploit the perennial availability of fruit in tropical forests. They are unrelated yet, because of their adaptations to eating fruit in different parts of the world, they have superficially similar bills which they use to reach fruit on thin branches that would not support their weight (Fig. 12). Another group of fruit-eaters, the fruit-bats (Pteropodidae) has evolved specialised palates containing ridges against which the tongue can crush fruit.

With special sources of food available all the year, in a constant climate, many tropical species have developed body structures which enable them to make the maximum use of a particular restricted diet. Hummingbirds (Trochilidae), hawk moths (Sphingidae), some sunbirds (Nectariniidae) and long-nosed bats (Glossophaginae) are not only specially equipped to eat nectar, but have also evolved the technique of hovering in front of flowers when feeding. This is achieved by rapid beating of the wings – more than 50 times a second in the case of humming birds. Bats of the genera *Glossophaga, Choeronycteris* and

Fig. 12 Beaks of toucan and hornbill (right), showing parallel adaptations for reaching fruit on thin branches.

probably, *Lonchophylla* alight on the corollas of large night-blooming flowers to lap, with their long tongues, the honeyed liquid at the bottom of the cup (Allen, 1939). In their search for food some have come to act as pollinators.

Arboreal Amphibia show many adaptations for life in the forest as do tree-dwelling reptiles such as geckos, snakes and so on. Many of them, brilliantly green in colour, are quite inconspicuous among the foliage. Others show various cryptic patterns.

The neotropical forest, in particular, is the great metropolis of tree-living termites, an important food source for many animals. Perhaps the heavy rains and consequent saturation of the soil may have driven these insects upwards, away from the floods. Certainly, ground-living species are normally found more commonly in drier savannah woodlands. But this may also be a consequence of the fact that climatic conditions above ground are too severe for arboreal termites to survive except in rain-forest. In rain-forest the mounds of the African *Macrotermes bellicosus* are broad based and only a few feet high whereas, in savannah, this species produces tall, steeple-shaped mounds which are usually several feet high (Olaniyan, 1968). *Cubitermes* spp. also build mounds while different species of *Nasutitermes* may construct either mounds or arboreal nests.

Termite nests are found in all strata of the rain-forest. They are made by cementing together particles of soil or wood and are often connected to the ground by covered passages through which the termites travel. The nests of some species which inhabit particularly wet regions are equipped with ridges which carry off excess rain water (Hesse, Allee and Schmidt, 1951). Termites are especially important in the ecology of the forest because their wood-eating habits hasten the decay of fallen trees. Indeed, their role in the production of humus parallels that of earthworms in temperate regions.

Tree-living animals not only possess the adaptations for climbing and gliding discussed in Chapter 13, but require good vision for accurate judging of distances when jumping or swinging from branch to branch. Binocular vision is found among primates and cats. The East Indian long-nosed tree snake *(Ahaetulla mycterizens)* has a long, narrow snout which allows the visual fields of its eyes to overlap, thus giving binocular vision. Other arboreal snakes and lizards rely on well-developed eyes rather than chemical senses to find their prey, for it is not easy to follow a scent trail through the branches of a tree.

Stratification. There is little doubt that, as knowledge increases, it will become possible to correlate animal communities with the different layers of the tropical rain-forest. Harrison (1957) divided the mammalian fauna of the rain-forest of Malaya into three zones as follows: canopy, 23 species; under-canopy, 23 species; ground, 51 species. When, however, he attempted to work backwords and define the layers represented by different species he found this classification to be unsatisfactory in two particulars. The group of 'under-canopy' animals, although representing a recognisable community, did not correspond to any layer that could be called an 'under-canopy'. Rather, they were animals with a vertical distribution to be found both in the canopy and on the ground. Secondly, the 'ground mammals' included two different elements – small animals such as rats which can climb well into the 'under-canopy' and larger species such as elephants and deer which have no climbing ability but crop leaves from bushes and small trees.

It was therefore necessary to re-define the mammals of the forest of Malaya and Australia as follows:
(i) *Upper air community:* Birds and bats which hunt above the canopy – mostly insectivorous but with a large proportion of carnivores;
(ii) *Canopy community:* Birds, fruit-bats, and other mammals confined to this zone – predominantly feeding on leaves, fruit or nectar, but with a few insectivorous and mixed feeders; (iii) *Middle-zone flying animals:* Birds and insectivorous bats – predominantly insectivorous, with a few carnivores; (iv) *Middle-zone scansorial animals:* Mammals which range up and down the trunks, entering both the canopy and ground zone – predominantly mixed feeders, with a few carnivores; (v) *Large ground animals:* Large mammals and, occasionally, birds living on the ground without climbing ability but of great range, either by reaching up into the cavity or by covering a large area of forest – plant feeders which mostly browse on leaves but sometimes eat fallen fruit (e.g. mouse-deer, *Tragulus* spp., and cassowaries, *Casuarius* spp.), or rooting for tubers etc. (e.g. pigs) with attendant large carnivores; (vi) *Small ground animals:* Birds and small mammals, capable of some climbing, which search the ground litter and the lower parts of tree-trunks for food – predominantly either insectivorous or mixed feeders, but with a fair proportion of vegetarians and some carnivores (Harrison, 1962).

A preliminary survey of the vertical stratification of small rodents was attempted by Southern and Hook (1963a) in forests close to the shores of Lake Victoria. Subsequently, in a nearby area of scrub regenerating to forest, Delany (1971) studied the occupation of niches by trapping both on and off the ground and by recording the plants from which animals were obtained. In this way some species were identified as ground-dwellers (e.g. *Lophuromys flavopunctatus* and *Hybomys univittatus),* some as inhabitants of trees and shrubs (e.g. *Thamnomys* spp. and *Praomys stella),* and others as commonly occurring in both situations (e.g. *Praomys morio* and *Oenomys hypoxanthus).* Although further information was obtained on the quality of the habitats in which the animals are found, other factors, as yet unknown, are clearly involved in their microdistribution (Delany, 1972).

The layered structure of the vegetation in the tropical rain-forest is mirrored in the stratification of bird life. Above the canopy live a few fast-flying species which feed on insects or other birds. Examples are offered by the Asian falconet *(Microhierax caerulescens)* and the black eagle *(Ictinaetus malayensis)* which lives on bats but also takes swiftlets, lizards and rats. Insects above the canopy are the prey of fast-flying swifts and swiftlets *(Chaetura* and *Collocalia* spp.*)*. Flight is restricted within the canopy where most of the bird life is found, including insectivores and fruit-eating species. The forest floor is the feeding ground of many large birds which rarely fly. These include the argus pheasant *(Argusianus argus)*, peacocks *(Pavo* spp.*)* and allied forms.

In some respects, the roof of the forest may be compared not only with the clearings but also with a savannah or prairie. There is the same wide green expanse, strewn with flowers and open to sun, rain and wind. Butterflies hover and grasshoppers skip over the surface whose denizens are exposed to the unrestricted view of predators (Haviland, 1926). In the upper strata of the Amazonian forest there are squirrels and sloths, tree-porcupines, tree-anteaters, tree-raccoons, wild cats and bands of monkeys. There are also innumerable birds – toucans, parakeets, barbets, frog-mouths, cotingas, pigeons, nightjars, curassows and bell-birds, tree-snakes, and tree-frogs. And there are vast hordes of insects: crickets, cicadas and other plant bugs, termites, ants, wasps, bees, beetles and flies that are never seen by man unless a tree is felled.

Protective devices of insects. Life is so intense and competitive in the rain-forest that adaptations to escape predatory enemies by utilising colour and pattern are more numerous than elsewhere. The abundance of protectively coloured insects is astounding. The number of species, especially of caterpillars and plant-bugs, which have evolved some form of terrifying device to bluff their enemies is likewise far greater than elsewhere. So also is the development of a nauseous taste combined with warning or aposematic colour to advertise unpalatability: and the tropical rain-forest is the chief home of mimicry in its various forms (Wells, Huxley and Wells, 1931).

Perhaps in response to the intense competition and heavy predation, many insect nests and cocoons are suspended by a thread in mid-air. Bates (1863) commented on the number of moth cocoons in the Amazon rain-forest and pointed out that these are well adapted to withstand the peck of a bird. Birds that are large enough to open the tough cocoons are either unable to obtain a purchase near enough, or, if they can do so, the cocoon simply swings away when pecked and so escapes damage. Belt (1874), however, considered this device to be a defence against marauding ants and remarked that, of all small animals, spiders in their webs stand the best chance of evading the attentions of hunting columns of army ants.

In order to understand the full significance of certain types of protective devices, a distinction must be made, as Hinton (1955) emphasised, between comparatively large vertebrate predators and small invertebrates (which are nearly always arthropods). Pensile cocoons are clearly a device that is chiefly effective against birds, as Bates realised. Nevertheless, they may in some instances serve as a protection against ants for, even if one should climb down the long suspending thread, it would probably be shaken off by the violent jerks of the pupa.

Some of the most remarkable pensile cocoons, according to Hinton (1955), are those of the Central and South American genus *Urodes. U. isthmiella* is one of the largest species and its obovate cocoons, 1.9 cm long, are suspended from a leaf by a silken cord over 32 cm long. The silk of both the cord and the cocoon is bright salmon red. The walls of the cocoon are an open meshwork with thick parallel threads about 1.5 mm apart and finer cross threads about 1 mm apart. The posterior end of the cocoon terminates in a short, open neck through which the larval cuticle is pushed out by the pupa, thus ensuring a clean, airy habitation free from anything that might become sodden and, by its smell, attract the attention of ants.

Solitary wasps construct exquisite little purse-shaped nests of vegetable fibre slung by a flexible stalk and stocked with insect prey. Most remarkable is one of the Chalcidae, an internal parasite of butterfly larvae which, when about to pupate, leaves the empty skin of its dead host and makes a cocoon the size of a cherry stone which dangles at the end of a thread up to 150 cm long (Haviland, 1926). South American wasps of the genus *Polybia* regularly nest in trees where ants of the genus *Azteca* also live (Jeanne, 1970), and some birds invariably build their pensile nests in trees protected both by ants and wasps (Myers, 1935).

Life at ground level

Those who come fresh to the rain-forest are generally disappointed at first by the paucity of animal life. This is partly because the fauna is dwarfed by the luxuriance of the vegetation and partly because the most abundant, active and brightly coloured species live in the upper storeys of the trees. One sees little of bird life in the Amazon forest largely, no doubt, because most species are hidden from view high up in the forest roof. Colonies of golden and black cassiques *(Cassicus persicus)*, however, which hang their bottle-shaped nests in the crowns of large trees such as the Brazil nut *(Bertholletia excelsa)* and silk cotton *(Bombex* sp.*)* are a glorious sight, as are the macaws, tanagers and humming-birds (Cott, 1930). Most animals are hidden behind the tangle of vines, trunks, branches and roots, or live unnoticed high in the forest canopy.

In general, the numbers of soil-dwelling invertebrates tend to be fewer in the tropics, and their sizes larger than those of their counterparts in temperate regions. This may be due to the rapid decay of organic matter and consequent absence of humus, lack of calcium owing to leaching, and the presence of very large numbers of ants and termites which consume large quantities of vegetable matter, but digest it so throughly that little is left over to enrich the earth (Cloudsley-Thompson, 1969). Invertebrates are generally well hidden. Worms, snails, millipedes, centipedes, woodlice, scorpions and insects retire under loose bark or decaying logs, or to the axils of palms. Land planarians crawl into the ground while land leeches climb the bushes from whence they attach themselves to passing mammals whose blood they suck (Wallace, 1869).

In addition to termites and ants, Coleoptera and Diptera are well represented. Amphibia are common on the forest floor, especially near streams and rivers. Reptiles are represented by monitor lizards, tortoises and venomous ground-living snakes such as the Gaboon viper *(Bitis gabonica)* and black cobra *(Naja melanoleuca)* of West Africa.

A few birds feed almost exclusively on the forest floor and also build their nests on the ground although, of course, in moments of danger they may take refuge by flying up into the trees. Notable examples from Africa include the crested guinea-fowl *(Guttera edouardi)* and a francolin *(Francolinus lathani)*. Several species of ground thrush *(Geokichla* spp.*)* are also essentially birds of the forest floor which provides their source of food, but their nests are found among the trees (Olaniyan, 1968).

With the exception of apes and squirrels, forest mammals are mostly nocturnal in habit and hide during the day. Ground-dwelling species tend to be small, stealthy forms that wind their way unobtrusively through the vegetation. Only the elephant moves by sheer strength, forming trails that are subsequently made use of by smaller forms including hippopotamuses, rhinos, buffaloes, pigs, tapirs and carnivores such as leopards and jaguars (Mertens, 1948).

The leafy canopy casts a heavy shade which inhibits grass from growing in the denser types of forest and therefore limits grass-eating animals to less dense or non-forested regions. Small forest browsers such as the South American red-rumped agouti *(Dasyprocta aguti)*, the Malayan chevrotain *(Tragulus javanicus)* and the African duiker *(Cephalophus maxwelli)* have all developed similar, small and compact shapes which allow them to run freely through the undergrowth. As the tree canopy develops more storeys, so the forest floor becomes more sparsely covered and more accessible to larger mammals.

The denser the forest, the less that air-currents can penetrate into it. Wind may be entirely absent among tall, closely packed trees. Mammals, therefore, tend to depend upon the sense of hearing rather than on sight or smell. During the day the forest is normally quiet but, with the approach of darkness, various Orthoptera burst into song, tree-frogs join in, and flocks of parrots and parakeets settle noisily into their nesting places. In America, the voices of howler monkeys add to the uproar (Hesse, Allee and Schmidt, 1951).

The number of species of large mammals is much smaller than in more open country. Their size is also less than that of their relatives outside the rain-forest. Forest elephants *(Loxodonta africana cyclotes)* are smaller than bush elephants *(L. africana africana).* They have small tusks and rounded ears. The forest buffaloes *(Syncerus caffer)* are smaller than the plains sub-species, and the same applies to leopards *(Panthera pardus).* The pigmy hippopotamus *(Choeropsis liberiensis)* like human pygmies, is a forest dweller, as are the small goat-antelopes *(Nemorhaedus* spp.*)* of Malaya. The royal antelope *(Neotragus pygmaeus)* of the coastal forests of West Africa is the smallest ruminant in the world, and many of the dwarf antelopes, such as *Hylamus batesi* of the Cameroons and *H. harrisoni* of the Semliki forest, are not much larger (Cloudsley-Thompson, 1969; 1972a). The forest-hog *(Hylochoerus meinertzhageni)* and bongo *(Boocercus euryceros)* would appear to be exceptional in their large size.

Symbiosis

The rain-forest is noteworthy for the number of symbiotic associations found in it between plants and animals. The three-toed sloth *(Bradypus tridactylus)* of South America provides an especially good example of the subtlety of such relationships between plants and animals. First, the sloth depends for its food upon the leaves of the *Cecropia* trees (Uritcaceae) which are characteristic of gaps and clearings in the forest. Sloths cling to the branches with their long, curved claws. They often hang motionless for long periods and move extremely slowly. They are then easily overlooked because their coats are a curious greenish-grey colour which makes them look more like ants' nests or masses of foliage than living animals. The long, coarse hairs of their fur are grooved and, in the grooves, live algae which give the sloths their green colour. Thereby they frequently escape notice from their major enemies, eagles and jaguars. But the interdependence of plants and animals does not end here. A species of moth is often found among the sloth's hair, where it lays its eggs: and the larvae of the moth in turn feed exclusively on the algae which grow only on the hair of the sloth (Richards, 1970).

Another example of symbiosis between animals and plants is afforded by myrmecophilous plants. Many species of *Cecropia* and *Triplaris* in the tropics have hollow trunks in which ants make their nests. The hollow petioles of *Tococa* spp., the hollow thorns of *Acacia* spp., and the bulbs of *Lecanopteris, Hydnophytum* and *Myrmecodia* spp. present cavities so admirably suited to the accommodation of ant colonies as to appear to have evolved for this specific function. Other rain-forest plants, such as epiphytes of the genus *Tillandsia,* are often inhabited by colonies of ants which nest in the spaces enclosed by their overlapping leaves (Wheeler, 1910).

Many myrmecophilous plants possess nectaries. On the leaf petioles of *Cecropia* spp., for example, are numerous minute structures (Mullerian bodies) containing oils and albumenoids. *Acacia sphaerocephala* possesses tiny succulent structures at the tips of the pinnules. These various organs provide food which attracts ants. The advantages to plants of being protected by the ants from animals that attack their foliage is said to have led by selection to the development of the special organs. Thus, in Brazil, ants of the genus *Azteca* to some extent protect *Cecropia lyratiloba* from ants of the genus *Atta* – these leaf-cutters sometimes despoil whole trees of their foliage. This interpretation, however, is opposed by workers who have observed the ants at first hand (Caullery, 1952; Janzen, 1967; 1973). Wheeler (1910) points out that, although *Cecropia peltata* resembles *C. lyratiloba* and *C. adenopus* in having Mullerian bodies, it thrives in Puerto Rico where no species of *Azteca* occur, and the tree is never inhabited by ants. Indeed, it flourishes on the mountains of the island even when its foliage is much eroded and perforated by insects. It is also claimed that even if *Atta* spp. are driven off when they invade trees inhabited by *Azteca* spp., the protection thus afforded is only incidental. To quote Ihering (1907), the *Cecropia* lives without *Azteca* as easily as a dog without fleas!

If ants are regarded as being entirely parasitic on myrmecophilous plants, however, it becomes impossible to explain the presence in these plants of the organs that provide food for the ants. There must, therefore, be some mutual advantage to the plant, thus making the relationship a symbiotic one. It should be remembered that a relationship need not be selected over a long period of time. In a similar way, Cott (1940) argues that adaptive coloration need only be of significance at one stage in the life of an animal in order to have been evolved by natural selection.

There is a close association between most epiphytes and ants. The latter sometimes live in special parts of the plants, such as the tubers of *Myrmecodia* and *Hydnophytum* spp. or in the sacs of *Dischidia* spp. They also inhabit the roots of epiphytic orchids. In such cases, the ants may assist the epiphytes by bringing back leaves, petals, seeds and other material from their foraging expeditions. This rots to form humus which provides the epiphytes with a source of moisture and nutrients (Richards, 1970).

The relations of ants to plants is discussed in great detail by Wheeler (1910). Here I will give only one more example – ant-gardens. This name was given by Ule (1904) to certain sponge-like ants' nests that he found on the branches of trees in the Amazon forest. These nests consist of soil carried up by various species of ants (*Azteca olithrix, A. ulei, A. traili* and *Camponotus femoratus*) and held together by the roots of numerous epiphytes which grow out on all sides. The ants not only perforate the soil with their galleries but, according to Ule, actually plant the epiphytes – a suggestion that Wheeler (1910) regards with scepticism.

Another group of ants, of special importance in rain-forest, includes the driver or army ants (Dorylinae), including the genera *Dorylus, Anomma* and *Aenictus* of the Old World, and *Eciton, Neivamyrmex, Labidus* and *Nomamyrmex* of the New. Their behaviour and social organisation has been described in detail by Schneirla (1971). These ants are nomadic: the colonies move from one bivouac area to another at fairly regular intervals. They send out massive predatory raids against soft-bodied and inactive insects which fall an easy prey to them. They also tear in pieces any other animal that is not quick enough to escape (Rettenmeyer, 1963).

Associated with raiding swarms of Neotropical ants are flies of the families Conopidae and Tachinidae which parasitise cockroaches and other insects disturbed by the ants (Rettenmeyer, 1961). Apparently these flies would be unable to find hosts without the aid of the ants, or would be much less successful at finding hosts.

African driver or safari ants also raid in large swarms and have many flies associated with their raids. Flies of the genus *Bengalia* (Calliphoridae), in particular, have frequently been recorded in association with ants or termites (Seguy, 1950). The adults are noticeable for their silent flights, the habit of pouncing on ants carrying larvae which are snatched and eaten (Cloudsley-Thompson, 1963a), and of sucking termites (Kemp, 1955) and ants. Correlated with such habits is their strong raptatorial and sucking proboscis (Fig. 13).

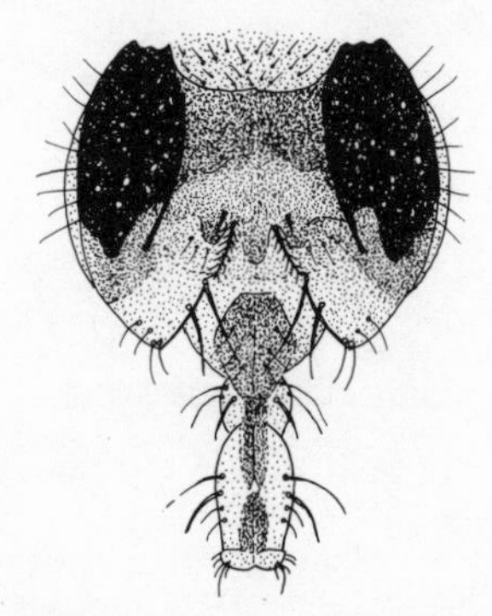

Fig. 13 Mouthparts of *Bengalia peuhi* (After Cloudsley-Thompson 1963a).

Both African and American doryline ants have large numbers of phorid flies living in their colonies, and a few species of Sphaeroceridae have been described as myrmecophiles with *Dorylus* spp. (Rettenmeyer, 1961). Thorpe (1942) observed *Stomoxys ochrosoma* hovering over columns of safari ants in East Africa and dropping eggs or larvae which were carried away by the workers. The relationships between flies and army ants probably range from symbiosis and commensalism to social parasitism and predation. In few cases have they been clearly defined.

4 SAVANNAH

Park-like savannah-woodlands are often found where the dry season is longer and the rainfall less heavy than in true closed forest. They separate the earth's belts of tropical forest from its desert regions, forming a wide range of transitional zones where moist and dry climates grade into one another and rainfall is often erratic. The trees tend to be widely scattered, except in favourable situations such as water courses. They show drought resistance and are often leafless during the dry season. In some regions they are replaced entirely by grasses.

Wooded savannah grades into tropical grassland steppe or savannah, which is dominated by Gramineae, although trees and bushes may still occur in open formation. There is usually a delicately balanced interaction between climate, soil, vegetation, animals and fire (Beard, 1953). In East Africa and elsewhere, the savannah has long been maintained by regular burning of the grass. When burning is prevented, the density of trees increases. Two main categories of grassland can be recognised: the edaphic* grasslands of the drainage lines, and the secondary grasslands which are mainly conditioned by manmade fires. The former sometimes degenerate into bogs and swamps which remain green well into the dry season. Consequently they are not usually burnt, a dense mat accumulates and soil moisture is conserved (Vesey-Fitzgerald, 1963). Thorn woodlands are even more xerophilous: switch plants with woody, photosynthetic stems, and succulents are characteristic. Grasses tend to be absent, or aggregate into clumps separated by bare soil.

Tropical savannah of various types is widely distributed throughout central and eastern Africa, north and south of the equator. In arid parts of India, human activities have created a short-grass savannah which is swept seasonally by the monsoon rains. Savannah is also found in South America, north and south of the Amazon rain-forest. The compos of Brazil and the llanos of Venezuela and Colombia consist of savannah grassland broken up by open forest. At the present time there is no consensus of opinion regarding either the origins or the classification of savannahs. In its broad sense the term is used to refer to a tropical formation where the grass stratum is continuous and important, but occasionally interrupted by trees and shrub. The stratum is burned from time to time, and the main growth patterns are closely associated with alternating wet and dry seasons (Boulière and Hadley, 1970).

---

* Edaphic – influenced by conditions of the soil and substratum.

Climate

Savannah occurs in tropical regions that experience a distinct dry season. Variability in precipitation is a distinctive feature of sub-humid and dry tropical climates. The rainy season is usually ushered in and out by violent thunderstorms and wind squalls, alternating with extremely hot sunshine. As the sun's vertical rays move northward from the equator, after the spring equinox, wind and pressure belts shift in the same direction, although lagging a month or two behind the sun. The inter-tropical convergence belt of heavy rains then creeps gradually northward and tornadoes or thunderstorms begin to appear in March or April over the savannah of Africa north of the equator. Rainfall continues to increase until July, or even August, when the inter-tropical convergence belt reaches its maximum northward migration. The rains decline as the belt retreats southwards, following the sun. By October or November, the dry, subsiding trades again prevail and drought grips the land (Trewartha, 1968). Similar conditions obtain during summer in the llanos of Venezuela and in the Indian savannah during the height of the monsoon.

There are two fundamental types of tropical climate – marine and continental (Appendix 1). The former does not show a pronounced dry season. It is restricted to a narrow strip on the eastern margins of the continents, but extends for some distance beyond the swing of the equatorial rains. Rainfall decreases steadily to the west, however, and in this direction forest passes gradually into savannah and desert. Tropical marine climates resemble equatorial climates in temperature, as in rainfall, and engender tropical rain-forest. It is the continental tropical climates that are largely responsible for savannah.

The duration and amount of rain decreases steadily in a poleward direction: while 130 cm (50 in.) may fall annually on the equatorial margin, this drops to 25 cm (10 in.) on the fringes of the desert. Although 50-100 cm (20-40 in.) is a usual amount, there is considerable variation from one year to another.

The campos of Brazil south of the equator has an annual rainfall from 100 cm (40 in.) to over 150 cm (60 in.), concentrated in the summer when the region is dominated by a northwesterly flow of moist, unstable equatorial air from the Amazon forest. The smaller llanos of Venezuela and Colombia forms the counterpart of the campos in the northern hemisphere. Its climate is similar, with a drought period extending to six or seven months of the year and accompanied by desiccating winds.

Vegetation

Savannah-woodlands. The vegetation of savannah-woodland is open;

rich in terrestrial herbs and grasses, but poor in lianas and epiphytes. The trees may be quite dense but, more usually, they are scattered and often stunted so that they may even be overtopped by tall grasses during the rainy season. By contrast with the uniformity of the equatorial régime, tropical continental climates are marked by seasonal rhythm. The vegetation grows with astonishing rapidity in wet weather and remains practically dormant during the dry season. Bulbous and other geophytes are often abundant.

The trees of the savannah are not forest trees. They are usually markedly xerophilous, hard and thorny, often with reduced leaves (e.g. *Acacia* spp.) or devices for the storage of water as in the baobab or 'tebaldi' tree (*Adansonia digitata*). Most of them are deciduous, shedding their leaves at the onset of the dry season and remaining without foliage until the return of the rains.

In some places as, for instance, South Africa, it is thought that the open 'parkland' or 'tree-veld' may be successional, the scattered *Acacia* spp. and other trees attracting birds and mammals that scatter their seeds. In this manner, too, a variety of other plants is introduced, so that a patchy type of woodland develops. Even this is probably not a true climax, however, but due to edaphic and biotic influences of which fire is the most important. Indeed, many trees become selected for fire-resistance (Polunin, 1960).

Orchard-type savannahs that have not been grossly overgrazed by domestic stock or concentrations of game in national parks, often support a rich and diverse vegetation. In Africa, forest trees, such as *Terminalia superba* and *T. ivorensis* are interspersed with shea butter (*Butyrospermum parkii*), palms (*Borassus aethiopum*), tussocky, fire-resistant grasses of the genera *Andropogon, Hyparrhenia* and *Pennisetum*. Some of the drier savannah belts are more heavily populated than those nearer the fringe of the rain-forest. This may have been due to the fact that the natural vegetation was not so dense as to make agriculture with primitive instruments too difficult. Crops such as millet, guinea corn and yams are grown. The land also supports large numbers of cattle and other domestic animals in Africa, because it is not too much infested by tsetse. Donkeys are very characteristic (Lawson, 1966). The failure of the East African groundnut project started after World War II highlights the danger of relying on fickle savannah rainfall for large-scale, mechanised, agricultural undertakings. *Thorn woodlands,* or tropical thorn forests, are usually even more erophilous. They are found chiefly where the annual rainfall is 40-90 cm (15-35 in.) and the mean temperature, high throughout the year ranges from 15-35°C (60-95°F). The foliage of the dominant trees is deciduous and often reduced. Switch-plants with woody, photosynthetic stems are also characteristic of the plant community. Many plants store water for the dry season in swollen trunks or roots. There are also sometimes tracts dominated by arborescent succulents such as the giant *Euphorbia* spp. of the Old World and the characteristic Cactaceae of the New World (Polunin, 1960).

Grassland savannah. Grassland constitutes one of the main vegetational types in the tropics and sub-tropics, but trees or tall bushes may occur in open formation and give a particular character to the landscape. Savannahs often cover vast tracts, though not without considerable local variation as in the cases of many South American campos and llanos.

The trees which appear at greater or lesser intervals are usually stunted and gnarled, but sometimes lofty. Many are deciduous but some, such as palms, are evergreen. The lower branches of the trees are often entangled with coarse and stiff grasses. Elephant grass *(Pennisetum purpureum)* may exceed 5 m (16 ft.) in height and form impenetrable thickets (Polunin, 1960).

In many areas of Africa, grassland savannah is maintained by regular burning or the intensive grazing of herds of wild or domesticated mammals. The former include many of the world's largest terrestrial animals. Where overgrazing and tramping by domestic stock are carried to extremes, erosion may lead to desert conditions.

The savannah of tropical America is a plant formation comprising a virtually continuous stratum of more or less xeromorphic herbs. Grasses and sedges are the principal components, with scattered shrubs, trees or palms sometimes present. It has been described in detail by Beard (1953) who recognises three sub-formations: tall bunch-grass savannah, short bunch-grass savannah and sedge savannah.

Savannahs occur under a great variety of climatic conditions from annual rainfall of 50 cm (20 in.) with seven or eight months of drought, up to 255 cm (100 in.) with negligible drought periods. Short bunch-grass savannah tends to predominate where there is less than 90 cm (35 in.) annually, and sedge savannah with over 200 cm (80 in.). All types of climate in lowland tropical America can support forest if conditions are favourable. Savannahs are found on badly drained land, with little relief, such as old alluvial plains or reduced uplands. They may be swept by regular fires and, except for herbs, the vegetation is so adapted as to be fire resistant. Nevertheless, grassland savannah is an edaphic climax and is determined by conditions of soil and locality.

The African savannah can be sub-divided into a number of distinct regions (Lawson, 1966). The most northerly of these, which grades into the southern border of the Saharan desert steppe, is known as 'Sahel savannah'. It enjoys a rainfall of about 25-50 cm (10-20 in.) concentrated in four to five months of the year. The vegetation is mostly thornland type. The trees are small, usually less than ten metres high, thorny and with narrow leaves. The grasses are sparse and seldom exceed one metre in height so that fires tend not to be very fierce or extensive.

The Sahel savannah is one of several long and comparatively narrow belts of vegetation that run across the African continent roughly parallel with the equator. On its southern side lies a zone of vegetation known as 'Sudan savannah' where the annual rainfall is in the region of 50-100 cm (20-40 in.) and the dry season lasts from October to April. *Acacia* trees are larger and more abundant than in Sahel savannah. *Acacia seyal* is typical, also the branching dôm palm *(Hyphaene thebaica)* whose leaves rattle noisily in the wind, and the massive baobab or tebaldi *(Adansonia digitata),* the trunks of which are hollowed out to store water during the dry season by the people of Darfur. Trees of the Sudan savannah are mostly deciduous. The herb stratum is continuous and dominated by perennial grasses from 1.0-1.5 m high. Grass fires are fiercer and more frequent than in the Sahel zone.

In northern Nigeria, the abundance of the most frequent trees, *Anogeissus leiocarpus, Combretum glutinosum* and *Strychnos spinosa,* is affected by soil conditions as well as by biotic factors. Thus, the first of these species is favoured by soils with a layer of clay-rubble that retains water; the other two trees dominate on drier, sandy soils (Keay, 1949). All the better soils, however, are heavily cultivated and the vegetation on them has been reduced to a uniform parkland which shows no significant variation between different soil types. The shrub savannahs of uncultivatable soils are likewise rather homogeneous (Clayton, 1963).

'Guinea savannah', which adjoins the rain-forest regions nearer the equator, is relatively moist, with a rainfall of 100-150 cm (40-60 in.) per annum, nearly all of which falls in seven or eight months of the year. The trees are mostly broad-leaved but lose their leaves for only a short period of the year. Usually up to about 18 m high, they may reach 30 m or more. *Isoberlinia doka* and *I. dalzielli* are the dominant species in the north, in association with very tall, fire-resistant grasses of the genera *Andropogon, Hyparrhenia* and *Pennisetum* as already mentioned. Big evergreen trees flourish in river valleys. The balance between trees and grass is a delicate one, maintained by annual fires whose flames sometimes reach to a height of 10 m. The southern type of Guinea savannah is characterised by such species as *Lophira lanceolata, Monotes kerstingii* and *Daniellia oliveri* but, of course, many local variations exist, often greatly influenced by human activities. Indeed, I believe that all the African savannah belts have been profoundly modified by man.

Although the main savannah areas lie to the north of the rain-forest, there are occasional patches of 'coastal savannah' in West Africa. For example, the coastal scrub and grassland zone of Ghana extends between the sea and the inland tropical forest. The factors that cause this rather unusual vegetation pattern are not fully understood, but it is apparently due to reduced rainfall within a narrow belt 24-32 km (15-20 miles) wide of the coastal land. It is likely that the combination of the angle of the coast, the direction of the monsoon and the distance from the shore of the Guinea current cause a reduction in rainfall. By contrast, the coastal region on the east side of the savannah belt is more humid than the inland country, forming the coastal forest-savannah of eastern Africa.

The northern areas of Guinea savannah, characterised by *I. doka,* are very similar to the 'Miombo savannah' which stretches across Africa from coast to coast south of the equator, from Angola to Mozambique. This is one of the great areas of relatively unspoiled wilderness that remain, because its soil is infertile and unsuitable for agriculture while tsetse excludes domestic stock and consequent overgrazing. The Miombo is florally richer than the Guinea savannah, probably because it occupies a larger and especially a much broader area so that more species have evolved in it. The dominant trees belong to the genera *Brachystegia* and *Julbernadia* which is closely related to *Isoberlinia.* In season, they are clothed in bright copper or red foliage that turns bronze, olive and finally bright green. Where the trees are dense, they suppress the grass beneath them so that fires are less fierce, and woodland dominates.

Natural drainage is perhaps the most important characteristic of tropical soils affecting the distribution of vegetation. Savannah vegetation characterises highly mature soils, either of senile or of very new land-forms, which are subjected to bad drainage and have intermittent perched water-tables so that there is an alternation of waterlogging with severe drought. The soils of upland forests are well-drained, either by virtue of their porosity, their relief, or both.

Fauna

While the fauna of rain-forest is remarkable chiefly for the extraordinary wealth and variety of species, animal life in savannah is characterised by the immense number of individuals of certain forms. Open country is a far less complex environment than rain-forest, and the number of ecological niches within it is much lower. It is exposed more to the influences of sunlight and wind, and climatic fluctuations are not buffered to the same extent as in forest.

Insects are often present in enormous numbers: three orders predominate. These are Orthoptera (locusts and grasshoppers), Isoptera (termites) and Hymenoptera (ants), but many others are well represented. Collembola, spiders and scorpions are also important. Termites deserve special mention. Typical of the African savannah are the snouted harvesters *(Trinervitermes* spp.*)* which are most abundant in this biome. Their mounds are small, but occur in large numbers. The workers collect grass at night and plug the exits to the mounds after the night's work. The jaws of the soldiers are greatly reduced and instead, they rely on chemical warfare for the defence of their colony. Their swollen heads contain a large gland which secretes a colourless irritant fluid. This is squirted at the enemy as a sticky thread from a pore at the end of the snout. An interesting behavioural adaptation for life in savannah is their habit of moving into underground galleries during the dry season and thereby avoiding the risk of being burnt during grass fires (Olaniyan, 1968).

Termites form a high proportion of the macrofauna of the soil in woodland savannah. They destroy growing plants, especially in times of drought, and thus prevent re-establishment of woodland; they modify the structure of the soil, and their mounds have topographical effects. Parkland savannah may sometimes consist of sharply delimited islands of woodland situated in grassland that is otherwise more or less devoid of trees. These islands are based on large mounds of *Macrotermes* spp. and the vegetation in them consists mainly of gallery-forest species. Such mounds in *Brachystegia*-woodland carry a specialized flora, mainly evergreen, with tall trees, whereas the rest of the flora is deciduous. Such mounds may even coalesce to produce closed canopy forest. Certain trees and shrubs (e.g. *Rhus, Grewia* and *Olea* spp.) are resistant to termite attack, and are often associated with termite mounds, especially in the Sudan savannah zone. Protected from fire by the absence of grass near the nests of the termites, they tend to form quite dense thickets. The fertile soil on the sides of termite mounds are often selected for cultivation by African farmers.

Amphibians are not common except near permanent water where they can breed, but reptiles are plentiful. Lizards, snakes and even tortoises are characteristic. Birds are common because food is plentiful and their mobility enables them to avoid conditions that are temporarily unfavourable. The dominant forms are ground living species such as ostriches *(Struthio camelus)* and emus *(Dromaius novaehollandiae)*, coursers and pratincoles (Glareolidae), bustards (Otididae), game-birds (Phasianidae) and sand-grouse (Pteroclidae). Predatory and scavenging birds such as eagles (Aquilidae), falcons (Falconidae) and vultures (Aegypiidae) thrive in the savannah on account of the good visibility (Cloudsley-Thompson, 1969).

The savannah is occupied not only by a population of sedentary birds but also, during the year, by a number of migratory forms. At least 135 Palaearctic species of birds migrate into Nigeria (Elgood, Sharland and Ward, 1966). In addition to the usual swallows and martins (Hirundinidae), shrikes (Laniidae), warblers (Sylviidae) and wheatears (Turdidae) (Moreau, 1967), a number of Palaearctic species of water-birds belonging to the families of ducks (Anatidae), herons (Ardeidae), plovers (Charadriidae), storks (Ciconiidae), cranes (Gruidae), gulls,

(Laridae), rails (Rallidae), avocets and stilts (Recurvirostridae), waders (Scolopacidae), and ibises and spoonbills (Threskiornithidae) spend their winters in the savannah south of the Sahara, across which they migrate. A high proportion of successful crossings shows how extremely well-adapted most species must be to undertake this journey, one of the most exacting in the world (Moreau, 1961).

Bird life of the Neotropical campos is also striking. Birds of prey, such as the buzzard *(Buteo magnirostris)*, hawks *(Buteogallus zonura* and *Rostrhamus* spp.*)* and the goshawk *(Accipiter tinus)* form a high percentage of the total species. Turkey vultures *(Catharistes urubu)* share with myriads of small red ants the important work of scavenging. They are to be seen soaring gracefully overhead or feasting on the carcasses of cattle. Several species of toucans (Ramphastidae) hide in the scattered trees and are rarely visible except when on the wing. Cuckoos *(Guira guira* and *Crotophaga ani)* are plentiful, as are parrots *(Chrysotis amazonica)*, nightjars *(Podager* sp.*)*, kingfishers (Alcedinidae) and snake-birds *(Plotus anhinger)*. Doves *(Columbula* spp.*)*, finches (Fringillidae) and woodpeckers (Picidae) are present in infinite variety, while flocks of parroquets *(Conurus aureus)*, often numbering many hundreds of individuals, chatter and scream all day in the mango trees (Cott, 1930).

The mammals of savannah, like those of steppe grassland, tend either to be large cursorial forms (p. 193) such as buffaloes and bison (Bovidae) antelopes and equines, or else small burrowing insectivores and rodents. In Australasia these groups are represented by marsupials which show parallel evolution in these respects with placental mammals.

Biomass. More than 40 species of large vegetarian mammals inhabit the savannahs of Africa, not counting smaller antelopes and other widespread herbivores. In any one habitat, such as the bushy grasslands of Kenya or the wooded savannah of Tanzania, as many as fifteen or sixteen species of large game animals may be found together, and population densities are often greater than in any other terrestrial habitat.

Such concentration does not, however, lead to severe competition because each animal has its own food preferences. Species of large ungulates found together must have different ecological requirements, at any rate at certain times of the year. Six different factors have been shown to be responsible for the ecological separation of herbivorous mammals living in close contact with one another in Tanzania. These are as follows: (a) The occupation of different vegetation types and broad habitats. For example, Grant's gazelle *(Gazella g. granti)*, wildebeest *(Connochaetus taurinus)* and zebra *(Equus burchelli)* prefer open grassland; dikdik *(Rhynchotragus kirkii)* and lesser kudu *(Strepsiceros imberbis)* the more densely wooded areas. *Oryx beisa* are tolerant of very dry conditions and thereby probably avoid competition with other grass-eating plains species.

(b) The selection of different types of food. The food taken by most species is, however, dictated partly by its availability. This is particularly true of the elephant *(Loxodonta africana)* which is very adaptable.

(c) Species with overlapping food preferences, such as wildebeest and buffalo *(Syncerus caffer)*, occupy different habitats.

(d) Different species may occupy the same place at different seasons of the year, e.g. Grant's gazelle and wildebeest, zebra and impala *(Aepyceros melampus)*.

(e) They may feed from different levels of the vegetation, e.g. giraffe *(Giraffa camelopardalis)*, black rhino *(Diceros bicornis)* and dikdik. Although black rhino browse about a foot from the ground, dikdik enter the dense canopy of shrubs and get food that rhinos do not reach.

(f) They may occupy different dry season refuges when competition for food is greatest (Lamprey, 1963).

The herbivores of the savannah include the world's largest terrestrial mammals. Concealment is difficult in open country, where size and speed are a better defence. Nevertheless, many species have effective camouflage. Although the stripes of the zebra are sometimes said to make the animals conspicuous, they tend to merge together in the distance forming a lightish colour that is actually less conspicuous, for example, than the darker hues of wildebeest, topi or tiang (Cloudsley-Thompson, 1967b). In open country concealment is not of great significance, for the zebra is gregarious and any enemy will soon be spotted by one or other of the herd. In close country or among scattered trees, however, the zebra is quite inconspicuous. As Cowles (1959) points out, it is in conditions such as this that it is in the greatest danger and can best benefit from concealment (p. 203).

Good vision is especially important in open country, and the eyes of grazing animals are usually situated far behind the mouth so that their possessors can see over the top of the grass while they feed. The eyes of a horse or zebra, for instance, are set high and well apart in the front of the head, giving almost all-round vision. At the same time, the eyeball is shaped so that both distant and nearby objects can be in focus at the same time (Fig.14) (Matthews and Carrington, 1970).

The biology of carnivores is less well known than that of herbivores. Each species occupies a particular niche, and the dual role of scavenger-predator has the great advantage of opening to exploitation the widest possible variety of food resources (Estes, 1967). In general, carnivores tend to prey on herbivores that are approximately the same size as themselves, although many exceptions to this rule may be found.

The food of lions *(Panthera leo)* varies from place to place. In some parts of eastern Africa, zebra is their favourite prey, in others wildebeest; but large numbers of smaller antelope and sometimes giraffe are killed.

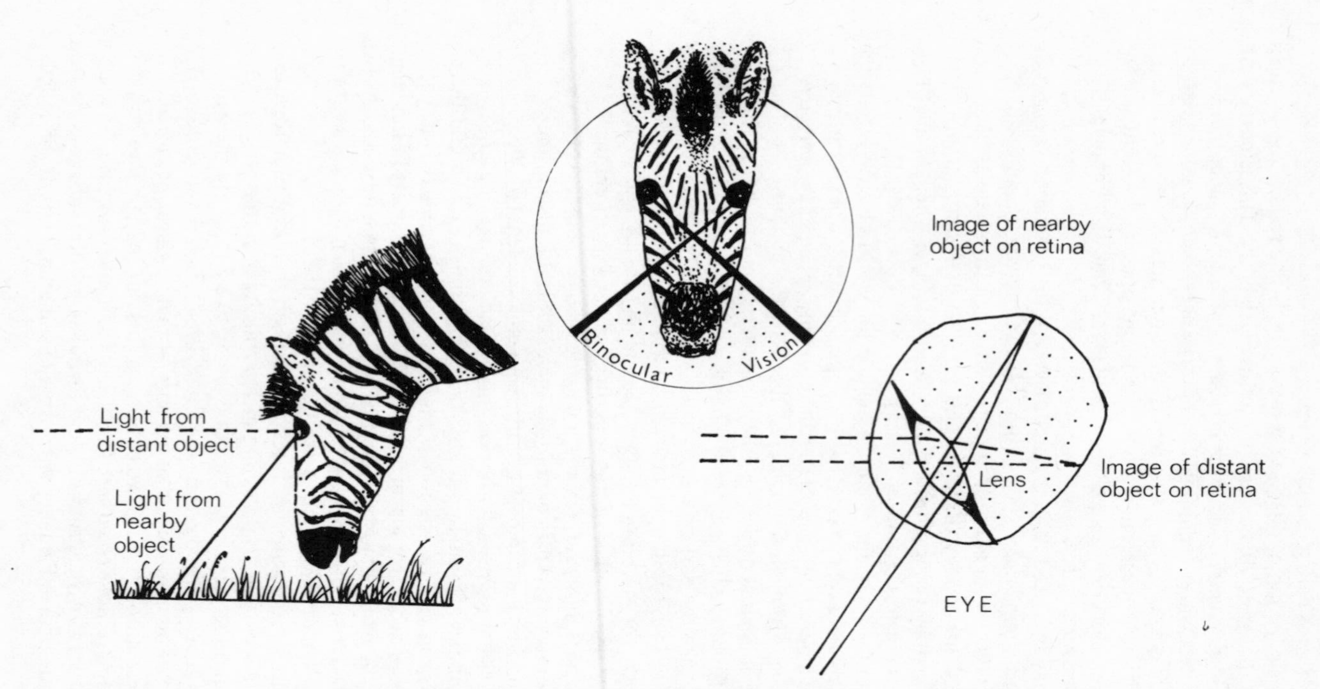

Fig. 14 Principles of vision in a grazing animal. The eyes of an equine enable it to keep distant and nearby objects in focus at the same time (After Matthews and Carrington, 1970).

In places where buffalo abound, lions feed largely upon them, and the art of tackling this formidable prey is handed down from old to young by imitation. Usually a single lioness takes the initiative in stalking and killing, sometimes assisted by other members of the pride. The male lion, unless alone, rarely hunts for himself, but he protects the pride's kill from thieving jackals and hyaenas (Estes, 1967).

Moreau (1969) has recently drawn attention to the deep dichotomy that exists in most groups of animals and plants between forest and non-forest species, at least in Africa. Even in animals as mobile as birds, it has been estimated that of 1,481 African species, less than 3 per cent are found in both types of environment (Moreau, 1966). In order to understand the functioning of the ecosystem and energy flow, trophic position and feeding habits as well as the distribution of species must be studied. It is probable that consumer populations are higher, show less variation and have a lower rate of turnover in unburnt savannah and in gallery forest than in burn savannah (Boulière and Hadley, 1970).

Seasonal rhythms

The lives of savannah animals are dominated by the seasons. After the rains, food is abundant, and herbivores disperse widely. In dry weather, however, many species migrate to areas near rivers and lakes where they can obtain drinking water and find better pasture. In the Rift Valley of Tanzania, for example, zebras and wildebeest migrate from places over 65 km away.

In the absence of marked changes in day length and temperature, the pronounced division of the savannah climate into wet and dry seasons is the main factor affecting the breeding of the fauna. The reproduction of birds is not regulated by photoperiod. Many African species breed during the short rains, but avoid the long rains (Moreau, 1950). The relationship between breeding and rainfall is nicely seen on the western shore of the Red Sea which receives winter rain. Here birds breed in winter, while inland they breed at the time of the summer rains. Thus, the white-bellied stork *(Sphenorynchos abdimii)* breeds in winter (December) in eastern Eritrea whereas, in western Eritrea at this time it migrates south, returning to breed only in summer (Smith, 1955).

In cases where breeding is inhibited when rainfall ceases, but evergreen vegetation remains unchanged, rain itself seems to be the proximate factor involved. Thus breeding is stimulated by rain in certain species of the Galapagos Islands (Lack, 1950); while *Geospiza* spp. brought back to California were stimulated to sing when rain fell on their cages (Orr, 1945).

Because much breeding takes place in anticipation of seasonal rains, but often when changes occur in the vegetation, it seems probable that such changes act as proximate releasers (Lofts and Murton, 1968). Thus, unseasonal breeding occurred in grass-warblers *(Cesticola* spp.*)* in Malawe when the flooding of banana plantations produced changes in the vegetation (Belcher, 1930). Marshall and Disney (1957) have demonstrated experimentally that the growth of green grass which occurs after rainfall is capable of stimulating the red-billed dioch *(Quelea quelea)* into breeding condition, and the same seems to be true of the blue-throated hummingbird *(Lampornis clemenciae)* according to Wagner (1959).

As far as ultimate factors are concerned, the breeding season of mammals may represent a compromise between the nutritional condition of the breeding pairs and the needs of the young for suitable food at critical periods of their growth. Likewise, in some species, their vulnerability may exert a selective pressure in favour of the young being born while there is adequate natural cover from enemies (Kingdon, 1971). The bi-annual breeding of the dikdik *(Rhynchotragus kirkii)* in Central Tanzania (Kellas, 1955) may be influenced by this factor, and the reproductive cycle of the gnu *(Connochaetes taurinus)* is probably largely shaped by predation (Estes, 1966). Thomson's gazelle *(Gazella thomsoni)* is another bi-annual breeder. At Serengeti, most of the young tend to be born at the beginning of the dry season in June, and another peak in births occurs in the short dry spell of January and February. Peaks in calving show significant local variation and, further North in Kenya, most fauns are born in the rains during April and November (Percival, 1928). Although such variations show no correlation with gross rainfall patterns, Brooks (1961) found a local correlation between births and fresh grazing. Thomson's gazelle is dependent on short grass and dry ground which must be found by seasonal migration, and the need for movement must subject the animal to pressures that do not operate on species that are more local in their habits. Whatever the ultimate factors in this breeding pattern, its rhythms presumably depend upon the females and young, since the males are continuously sexed (Kingdon, 1971).

By contrast, male topi *(Damaliscus korrigum)* have rutting peaks, which must largely determine the peaks in the numbers of births. In Serengeti these occur mostly in August – September and, in smaller numbers, in January-February. The same species has single peaks of calving in the Rukwa Valley (September) and in West Uganda (January-February). In all cases, young topi are born when the weather is relatively dry and the grass is short. They grow very quickly so that optimum feeding conditions for the growing calves may be an important ultimate factor in determining the seasonal periodicity (Kingdon, 1971).

Variations occur, however, between populations living under the same climatic conditions. For instance, Laws and Parker (1968) found differences in the breeding pattern of two elephant populations separated by the River Nile at Murchison Falls. The southern population is very dense and has a seasonal peak in conceptions which is five months later than that of the population north of the river, where conception is associated with the beginning of the rains. Higher densities and nutritional deficiences in the southern population are thought to be responsible for the discrepancy.

Several different methods have been used to assess the breeding condition of small mammals. These include fecundity of males and the frequency and size of placental scars, but the simplest and most common method involves determination of the percentage of adult females pregnant or lactating in each month of the year (Delany, 1972). When the reproductive behaviour of harsh-furred rats *(Lophuromys flavopunctatus)* is compared in various regions, it is found that peaks of breeding either coincide with the period of maximum rainfall, or else the maximum breeding precedes the rainfall peak by one month. Periods of minimal breeding, on the other hand, occur during the dry periods even in regions of moderately high rainfall (Dieterlen, 1967; Delany, 1971; Hanney, 1964).

Probably very few mammals do not to some extent change their feeding habits with the seasons. The arrival of the rains in drier savannah induces such an outburst of vegetation, insects and other animals, that the choice of diet is much increased. For instance, during the rains, half the food of the kudu *(Strepsiceros strepsiceros)* consists of succulents which are not eaten at all during the dry season (Harrison, 1936). Outbreaks of man-eating by lions *(Panthera leo)* are likewise seasonal. Along the coastal belt of Tanzania, bush pigs, wart hogs, and other small game that survive the expansion of the human population, form the staple food of lions. During the rains, when the animals are dispersed and difficult to find, hungry lions, and especially young animals that are just learning to fend for themselves, turn to humans as a tempting source of meat (Ionides, 1965).

The breeding of savannah reptiles and, even more, of Amphibia is very largely dependent on rainfall. The Nile crocodile *(Crocodylus niloticus)* lays its eggs during the dry season when water levels are falling. Incubation coincides with the phase of lowest water, and hatching occurs with the onset of the rains (Cott, 1961). Lizards also breed when the weather is wet: no doubt the presence of abundant insect food is an important ultimate factor. Numerous inter-related factors influence reproductive cycles. These may be determined by innate rhythms of the animals themselves, as in the copperhead *(Agkistrodon contortrix)* and certain cobras (Elapidae). In the Nearctic five-lined skink *(Eumeces fasciatus),* the breeding cycle seems to be triggered by a hibernation period. On emergence, the animal undergoes a rapid physiological change to prepare it for the season of reproduction (Fitch, 1954). It seems doubtful, however, that tropical reptiles either hibernate or aestivate: nor that their breeding seasons should be regulated by photoperiod (Fitch, 1970).

Seasonal fluctuations occur, not only in the numbers and composition of the invertebrate fauna of the savannah, but also in the time of activity of the various species – a trait even more conspicuous in deserts (p. 100).

Diurnal rhythms

Although not so marked as in desert regions, where the climatic changes between day and night are particularly great, diurnal rhythms tend to be more pronounced in savannah than in the dense shade of equatorial rain-forest. Some animals are nocturnal, others day active. Variations in light intensity act as synchronizers to keep the endogenous circadian rhythms of arthropods in phase with diurnal climatic changes. (Cloudsley-Thompson, 1961a). Thus they are the proximate factors that influence the physiology of the animals. What the ultimate factors may be is little known (Cloudsley-Thompson, 1960b; 1969). In some cases, these tend to be concerned with water conservation and thermo-regulation. In others, biotic factors are more important. The synchronization of sexual partners or the relationship between predators and prey may be as important as the selection of appropriate microclimatic conditions.

The adaptive significance to bats of emerging from their roosts each evening at exactly the same time in relation to sunset, probably lies in the avoidance of predators such as kites *(Milvus migrans)* and the bat-hawk *(Machaerhamphus anderssoni)* which hunt by sight, and in obtaining the greatest number of crepuscular insects. In Khartoum there is often a brief overlap between the roosting of the kites and the emergence of bats. During this period I have sometimes seen the earliest bats caught by the last of the kites retiring to their roosts for the night (Cloudsley-Thompson, 1970a).

Few quantitative studies have been made on the rhythmic behaviour of game animals, but censuses carried out at water holes in Rhodesia have provided data on the daily drinking habits of the common herbivores and carnivores. Elephant, buffalo, zebra and giraffe tend to drink in the evening and during the night; wildebeest, roan *(Hippotragus equinus)*, sable *(H. niger)* and kudu at night and in the morning. Wart-hogs *(Phacochoerus aethiopicus)* drink exclusively during the day, lion mostly during the day, leopards *(Panthera pardus)* and hyaenas *(Crocuta crocuta)* at night (Weir and Davidson, 1965).

Most savannah birds and reptiles are day-active; amphibians which, anyway, tend to be scarce away from water, are nocturnal. This is true of the African toad *(Bufo regularis)* (Cloudsley-Thompson, 1967c). The days are spent in hiding but, at dusk, there is a movement to or from pools or rivers (Chapman and Chapman, 1958). Diurnal rhythms of activity in reptiles are primarily concerned with thermoregulation by behavioural means (Cloudsley-Thompson, 1971a).

Cursorial life

A few savannah herbivores, such as the elephant, are large enough to be immune from the attacks of predators. Others, such as rodents, obtain protection by burrowing. Many, however, rely on speed to escape from their enemies. These show various adaptations to fast movement (discussed in Chapter 13) and include both mammals and cursorial birds.

Many of the species of animals that graze and browse in savannah live together in great herds. This conveys a certain amount of protection because a group of vigilant animals is more likely than a solitary individual to perceive the approach of a predator. Herds of elephant may number more than 100 individuals, buffalo herds between 500 and 600. The largest herds of all are the congregations of 10,000 or more wildebeest and zebras which live on the plains of East Africa (Matthews and Carrington, 1970).

While hoofed grazing mammals evolved in other continents, a remarkable array of hopping marsupials emerged in Australia. Kangaroos comprise some 50 species ranging in size from the red kangaroo (*Macropus rufus*) which stands about 1.5 m in its usual squatting position, to the diminuitive sombre wallaby (*M. browni*) which is only about 30 cm in height. Kangaroos have gained speed and mobility by evolving elongated hind legs and feet and adopting a bipedal, bounding gait. The long, muscular tail acts as a counterweight to the body and stabilizes the animal as it hops along. It acts as a prop when its possessor is at rest or grazing, and becomes a lever on which the animal rears its body when fighting (Matthews and Carrington, 1970).

The feet of kangaroos exhibit syndactyly, or fusion of the digits. This indicates an arboreal ancestry, although kangaroos have lost the climbing habit and their feet are not longer prehensile. Even tree-kangaroos (*Dendrolagus* spp.) are unable to grip branches with their feet and have to rely for security upon their claws and broadened soles. The opossum (*Didelphys azarae*) of the South American pampas and Andean forests climbs as nimbly as ever when trees are available.

5 DESERT

Beyond the limits of the swing of the equatorial rainfall belt, at the latitudes in which the trade winds blow throughout the year, lie the world's greatest deserts, where the annual precipitation is less than 25.5 cm (10 in.). Hot deserts, such as the Sahara and Kalahari, have no cold season but, in the so-called 'cold deserts' like the Gobi and Great Basin, one or more of the winter months has a mean temperature below 6°C (43°F).

The deserts of the world can be divided into five types on a climatic basis. These are as follows: sub-tropical deserts, cool coastal deserts, rain-shadow deserts, interior continental deserts and polar deserts. Only the first four of these (Appendix 2) will be discussed in the present chapter, but all of them are arid. In the case of the polar deserts (Chapter 9) water is only present in the form of ice and is therefore not available to plants and animals. In the other types of desert, water is deficient throughout most of the year because the amount of evaporation greatly exceeds the annual precipitation.

The distribution of deserts throughout the world is due mainly to the way in which the atmosphere circulates, particularly in its lower layers. Sub-tropical deserts are the result of semi-permanent belts of high pressure in tropical regions within which the air has a tendency to descend from high altitudes towards the surface of the land. At the beginning of its descent this air is cold and dry but it becomes warmed by compressional heating at the adiabatic rate of 10 degrees C per 1,000 m. Consequently it reaches ground level very hot and with an extremely low relative humidity so that it is completely incapable of producing any precipitation.

Cool coastal deserts are almost always rainless yet drenched with chilly moisture. They include the Namib, Atacama and the coastal desert of Baja California. Rainlessness results from descending air masses, the high humidity and cold from nearby cool ocean currents, respectively the Benguela, Humboldt and Californian Currents. In each case these currents have originated in polar regions, but they pull up even colder water from the depths which lie near the shore in those parts of the world.

Rain-shadow deserts are situated on the lee sides of mountains which cause the prevailing wind to rise and drop its moisture in the form of orographic precipitation. An example is afforded by the Mojave Desert which owes its winter aridity to the Sierrra Nevada and Transverse

Ranges, while its aridity in summer is caused by the presence at that time of the sub-tropical high pressure cell which dominates the Sonoran and Chihuahuan Deserts, just to the south, throughout the year. The Great Basin Desert is likewise sheltered by the Sierra and Cascade Ranges to the west and by the Rocky Mountains to the east. The deserts of Patagonia are in rain-shadow as far as the prevailing westerly winds from the Pacific are concerned, while air masses moving from the south Atlantic are cooled by the Falkland Islands Current and carry little moisture.

The Australian desert, too, is to some extent in the rain shadow of the Great Dividing Range. Like the Mojave and Great Basin Deserts, however, it also falls into the category of the interior continental deserts, which are arid through the lack of marine influence and other factors related to the massive bulk of the land surrounding them. Distance from water is the final factor in the creation of the deserts of central Asia (Logan, 1968; Wallén, 1966).

Evidence has recently accumulated which shows that most of the world's desert regions, including the vast Sahara (Cloudsley-Thompson, 1971c) are largely man-made and are expanding extremely rapidly. Between 1882 and 1952, land classified as desert or wasteland increased from 9.4 to 23.3 per cent of the total surface of the earth (Ehrlich and Ehrlich, 1970). Destruction of the vegetation is caused by felling trees for firewood, and overgrazing, especially by goats which can eke out a living where any other domestic animal would starve.

Climate

Desert regions are not necessarily characterised by great heat, nor do they always consist of vast expanses of shifting sand dunes. The one characteristic common to them all is their aridity throughout most or all of the year. When the air is humid, not only does less solar heat penetrate to the ground during the day, but less is lost from the earth by radiation at night. Consequently, humid climates tend to show daily or seasonal stability, while deserts are characterised by extremes of temperature and humidity. The most adverse conditions for life consist of a combination of aridity and high temperature, and it is the effects of these two factors that have been studied most extensively.

Weather or climate exist as part of a global system of air movements and desert areas are often small in relation to the wind systems that dominate them. A detailed bibliographic review of the state of knowledge about desert climates is given by Reitan and Green (1968).

Desert climates are subject to extremes. High temperatures and low humidity during the day are followed by comparatively cold nights. long periods of drought are broken by torrential rainfall and flooding.

Although desert rainfall tends to be seasonal, it is most erratic, and the total annual precipitation varies considerably from year to year. The presence of *wadis* and dry saline lake beds, such as the *chotts* of North Africa, show that torrential rain may sometimes fall. Most of this runs off so rapidly, however, that it tends to be wasted.

Whereas, in clear desert air, only about 10 per cent of solar radiation is deflected by dust particles and cloud, in humid regions some 20 per cent may be deflected by clouds, 10 per cent by dust and 30 per cent by water surfaces and vegetation. At night on the other hand, up to 90 per cent of accumulated heat escapes from the desert surface to the upper air while, in humid countries, only 50 per cent escapes, 30 per cent being deflected by clouds and dust, and the remainder being retained by the land cover and water. Consequently, humid equatorial climates tend to show diurnal and seasonal stability whereas deserts are characterised by extremes of temperature and humidity.

Maximum temperatures are high in deserts and semi-arid areas, especially in summer. Even in winter, temperatures during day-time are very high compared with other regions. The mean maximum temperature may reach 43.6°C in Baghdad, 41.2°C in Biskra and 40.0°C in Phoenix. In arid regions, influenced by cool ocean currents, mean maximum temperatures in summer are considerably lower – for example, 24.4°C in Antofagasta and 29.4°C in Windhoek. Winter mean maxima are low in the cold interior continental deserts of Asia. In Kashgar the mean maximum is only 0.6°C in January but it is as high as 30°in Khartoum and Bilma (Wallen, 1966).

Mean figures are of little biological significance in deserts, however, because fluctuations are so large. It is the incidence of high temperatures liable to cause heat damage and of low temperatures which cause frost damage and limit the growing season of the vegetation, that are important to plants and animals. Shade temperatures as high as 56.5°C have been registered in Death Valley, California, and even higher in the Sahara. An annual range of shade temperature from -2°C to 52.5°C has been recorded from Wadi Halfa and a daily range in summer of 29 deg.C at In Salah. The record appears to be a fluctuation of 38 deg.C (-0.5°C to 37.5°C) within 24 hours at Bir Mighla in southern Tripolitania in December. During 1910 there were 14 days of frost at Tamanrasset at an altitude of 1400 m, and absolute minima of -7°C and -2°C were recorded in January and February of that year (Cloudsley-Thompson and Chadwick, 1954).

Not only is lack of moisture in the form of rain the chief factor causing desert conditions, and the absence of clouds responsible for extremes of temperature, but low humidity itself has an adverse effect upon plant and animal life. This is because the saturation deficiency of the atmosphere increases as temperatures soar during the day.

Conversely, there are several examples in low latitudes where high relative humidity, usually due to dominant onshore winds from the sea, may to a certain degree compensate for lack of rainfall. In the arid and semi-arid zones along the Persian Gulf, for instance, the vegetation is much better developed than would be expected from the rainfall and even includes several species characteristic of a more humid tropical flora. Again, animal life in the Namib Desert depends for its moisture upon fog that comes in from the sea, while energy is supplied to the ecosystem by fragments of dry vegetation blown by the easterly berg winds of the region. The low humidity of many desert climates greatly affects the flora and fauna by influencing transpiration rates.

Strong winds and sandstorms are characteristic of desert climates, especially in summer. They blow hardest during the day, while the nights are relatively calm. Although speeds seldom exceed 80 km/hr and average only 16 km/hr annually, their effect is enhanced by low humidity, high temperature and lack of shelter.

Vegetation

Most desert and semi-desert regions support some vegetation although their climate and soils are dry. The least amount of plant life is found in areas of *hammada* or denuded rock, except where these are traversed by wadis. Dunes are usually quite bare but sandy desert, in general, has a less scanty flora. Spacing of desert vegetation reduces competition for scarce resources of water.

Beyond the edges of true desert are shrub-steppe lands where the rainfall is very scanty. Such is the *Acacia* desert-scrub which lies south of the Sahara, extending from Senegal to the Red Sea and the mountains of Ethiopia. Where rain is moderate in amount, even if it falls only in a few days of the year, grassland savannah is found.

Desert plants are adapted in various ways to withstand the adverse conditions under which they have to live. The ultimate stress suffered by desert plants is the dehydration of their protoplasm. The amount of water that can be lost at the cellular level before irreversible dehydration sets in varies greatly in different species. In the creosote bush *(Larrea divaricata),* the water content of the leaves may drop to 50 per cent of their dry weight but the plant will recover when adequate water is supplied. The water content of the leaves of forest trees is usually in the range 100-300 per cent of the leaves' dry weight, and varies very little.

Root systems may be widely spreading as in the saguaro cactus *(Carnegiea gigantea),* or deeply penetrating as in the mesquite *(Prosopis juliflora)* where they can reach a depth of over 50 m. Succulent plants, such as cacti and agaves, store water in roots, stems or leaves. The barrel cactus *(Ferocactus wislizeni)* stores so much that it has been used as an emergency source of water by Indians and other inhabitants of the desert. Water loss by transpiration may be reduced by evolving small, fleshy leaves, a thick, waxy cuticle or a downy, hairlike covering.

Xerophytes can be further divided into:
(a) Drought-escaping ephemerals, which germinate and flower rapidly after rainfall; (b) Drought-evading plants which are so small that they conserve the little water available through their restricted growth; (c) Drought-enduring species, like the creosote bush which ceases to grow when soil moisture is absent; and (d) Drought-resisting succulents with reserves of water in their tissues (Cloudsley-Thompson and Chadwick, 1964; Cloudsley-Thompson, 1965a).

Seed dispersal

Many species of drought-evading plant produce seeds equipped with dispersal units which aid their distribution. In this way, the offspring have an increased chance of reaching a site that is suitable for germination. Other species, however, have evolved hygrochastic mechanisms. Stems, fruit-stalks, valves and involucral bracts are closely curled towards one another when dry, but open apart when moistened. By this means seeds are scattered only during the moist season. This tends to prevent long distance dispersal of seeds from their parent plant which presumably is already growing in a favourable locality.

Aristida funiculata provides an excellent example of an annual grass with a well-developed dispersal mechanism. Dispersal consists in a large number of seeds becoming entwined to form a dense, rounded ball which is blown over the desert surface by the wind. As the sharp points come in contact with the soil, individual units become anchored and detached, until the ball finally disintegrates. Changes in the humidity of the air subsequently cause each seed to twist like a drill, forcing it even deeper into the soil where it remains until rain falls. Individual seeds may also become attached to the hairs of animals – the legs of goats are sometimes completely covered with them – and are thereby dispersed. Many other plants of desert regions have well-developed barbs and bristles to aid their dispersal by animals. Examples are afforded by the spikelets found in bur grasses of the genus *Cenchrus* as well as in *Setaria* and *Tragus* spp.

The fruits of *Colocynthus vulgaris* are spherical, thin-walled gourds, about the size of an orange. They are blown about the desert and semi-arid regions of the Sudan, sometimes for considerable distances until they become trapped against a rock or, more usually, in a hollow where eventually they disintegrate or become buried in wind blown sand. When rain falls and collects in the hollows, the seeds germinate in favourable surroundings. Other rolling plants of the desert include *Anastatica hierochuntica,* a crucifer, and *Asteriscus pygmaeus,* a composite, to both of which the name 'Rose of Jericho' is sometimes applied (Fig. 15).

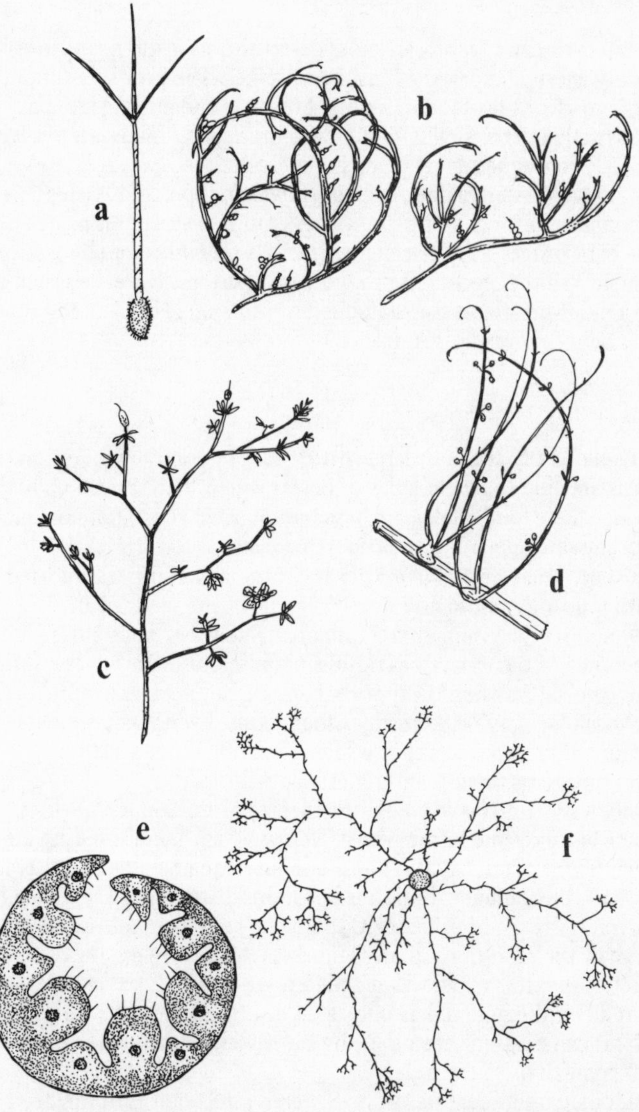

Fig. 15 Adaptations of desert plants: (a) Dispersal unit of grass *(Aristida)* (2cm); (b) 'Rose of Jericho' *(Anastatica),* dry (8cm) and wet (10cm); Creosote bush *(Larrea divaricata)* (30cm); (d) Aphyllous plant *(Calligonum comosum)* (25cm); (e) Rolled xerophytic grass leaf in section (2mm); (f) Plan of rooting system of a cactus (50cm). (Drawings not to scale.) (After Cloudsley-Thompson, 1965a).

A characteristic feature of most desert shrubs is the possession of painfully sharp and prickly thorns and spikes. The probable function of these is to afford protection against browsing and grazing animals. In the Australian deserts, where there is more vegetation as alternative food and, consequently, less grazing pressure, the *Acacias* are less thorny. Many desert plants, such as *Euphorbia* spp. and Asclepiadaceae also contain poisonous or irritant latex. Others secrete resins or tannins in the bark or leaves, whilst the pods of *Cassia senna* contain a strong purgative and the creosote bush has a pungent smell. Devices such as these render the plants unpalatable, both to animals and to insects (Cloudsley-Thompson, 1965a).

Fauna

The fauna of any region is largely dependent upon the vegetation. Some animals are able to live in areas of desert where there are no plants growing: their food chains are based upon dried vegetation and grass seeds, blown often from a considerable distance. As Brinck (1956) points out, dried vegetable matter is continually being transported by the wind into the more arid desert regions of the world. Part of this is eaten or destroyed immediately; the remainder becomes buried and, in the absence of bacteria, does not decompose but comes to the surface, often years later, when it supports Lepismidae, the larvae of Tenebrionidae, and occasionally other forms. Even vegetationless desert, therefore, may support a sparse fauna, provided that a sufficient concentration of dried plant material is achieved.

Only a short distance below the surface of the soil, conditions become less extreme in the desert. Below 50 cm there is hardly any diurnal temperature variation in the sands of the Sahara and at twice this depth the annual variation is not more than 10 deg.C (18 deg.F). A temperature range of 30.5 deg.C (55 deg.F) has been measured on the surface of the sand in southern Tunisia in April: five cm down a cricket's hole, this range was reduced to 18 deg.C (32.5 deg.F), while 30 cm down the hole it was only 12.5 deg.C (22.5 deg.F). From such results it can easily be seen that, by burrowing, an animal can avoid the worst conditions of the desert.

An outstanding feature of the Saharan sands that is of major importance to their inhabitants is their comparatively high humidity. Even during the summer, the air surrounding loose grains at a depth of 50 cm has a relative humidity of 50 per cent. This moisture rises from extra-Saharan water which underlies most of the dunes occurring, for example, at a depth of 5-25 metres in the Grand Erg Occidental. This is another advantage to burrowing animals because saturation deficiency, the evaporating power of the air, varies inversely with temperature so that extremes of both heat and drought are reduced underground (Pierre, 1958).

Caves and rock fissures form the natural habitat of many desert animals because the microclimatic conditions with them are relatively uniform. Consequently, within distances of a relatively few metres, there is available in arid country a very wide range of microclimatic conditions from among which an animal can choose the most suitable with little expenditure of energy.

Insects. Of the known orders of purely terrestrial insects, a high proportion is found in the desert. No less than 26 out of 32 are represented in the north-west Sahara. In absolute desert the number is reduced: only fourteen are found and, of these, not all are sufficiently numerous to merit consideration.

Thysanura or bristle-tails of the family Lepismatidae are abundant 'sabulicoles' or sand-dwellers, commonly found under rocks, dried dung and so on. They feed on dry vegetation and, in extremely arid regions such as the Namib desert, form the basis of a food chain on which many carnivorous animals depend (Lawrence, 1959).

Orthoptera, comprising crickets, locusts and grasshoppers, are an important element of the desert fauna and appear in abundance on the vegetation that springs up after rain. The desert locust *(Schistocerca gregaria)* exists in the Great Palaearctic Desert as a result of its migratory habits. Solitary locusts are normally found among sand dunes on coastal plains, scrub belts along the beds of wadis, oases, and similar habitats which represent ecological islands in the deserts.

The migratory form appears when population densities build up. Long distance movements of swarms take place high in the air where wind speeds are often greater than the speed of flight of the locusts. Consequently, it does not much matter in what direction the locusts are actually heading. Inevitably they get carried into areas of low barometric pressure where rain is most likely to fall. Here they feed on the ephemeral grasses that spring up and, later, deposit their eggs. The female locust prefers to lay in sand that is dry on the surface but damp underneath, and the eggs do not develop unless they are kept moist (Rainey, 1951).

Grasshoppers and crickets are common throughout the deserts of the world. Whereas grasshoppers show little adaptation to their environment apart from possessing desert-coloration, some crickets such as the sand-treaders *(Ammobaetes* and *Macrobaetes* spp.*)* of North America possess combs of long hairs on their lower hind legs which enable them to make sure progress in sand. They lie buried during the day and emerge at night, especially in warm weather.

Many species of Hemiptera (bugs) live on the sparse vegetation of desert regions and the most conspicuous family, the Cicadidae contains several sand-dwelling species. Coccidae are fairly numerous in desert areas: the 'manna' which sustained the Israelites during their wanderings in Sinai, was the secretion of *Trabutina mannipara. Margarodes vitium* which occurs in arid parts of South America becomes completely covered with a waxy coating in which condition it can resist prolonged drying (for up to seventeen years). When the encysted insect is put into damp soil, however, it absorbs moisture and continues its development.

Tropical termites usually extend their nests in a north-south direction so that a comparatively small area is exposed to the noon-day sun. Dead shrubs in the Sonoran desert are usually eaten away by *Amitermes arizonensis* while other species of the same genus are found living at the bases of agaves, ocotillo, cholla and other cactuses. In the more arid deserts, only subterranean termites occur. Although earwigs (Dermaptera) are nearly always found in moist situations in temperate regions, some of the larger species, such as the cosmopolitan *Labidura riparia,* are equally at home in quite arid places. *L. riparia* is a nocturnal carnivore and hides under stones and in cracks in the soil during the day.

Ant-lions (Neuroptera; Myrmeleontidae), whose larvae have been called 'demons of the dust', are characteristic insects of the summer season. Their cone-shaped pits can be seen especially well in fine sand. At the bottom of each pit lies buried an ant-lion larva, in ambush for passing ants and other insects that may fall into the trap. The prey is encouraged to slide to its doom by particles of sand flicked with surprising force and accuracy by the larva waiting below. In hot weather, these larvae orient themselves within their pits so that their bodies are in the coolest regions available to them (Cloudsley-Thompson and Chadwick, 1964).

The most conspicuous Hymenoptera of deserts are wasps and ants. Bees require nectar and pollen and are restricted to the rainy season. Most wasp species are burrowers and excavate holes and galleries in sandy soil where they deposit paralysed spiders and insects which serve as food for their larvae. The majority of ants live in subterranean nests although a few, such as *Crematogaster* and *Camponotus* spp. nest in the wood of dead trees or under bark.

Paradoxical as it may appear, the primary adaptations of ants to life in the desert are not related so much to the dryness of the air as to the nature of the terrain and the upkeep of their nests. Most of the species found in the Sahara Desert, for example, live in comparatively moist localities and only a few are found in really arid soils. The most resistant species *(Acantholepis frauenfeldi)* brings water to its nest from the salty, damp sand of water-bearing strata. In a similar way, the harvester ant *(Veromesser pergandei)* of California is susceptible to extremes of heat and cold, but lives in deep nests and forages only during a brief period each day when the temperature is favourable (Bernard, 1948).

Lepidoptera of the desert do not at first appear to show any special adaptations beyond pupal diapause which, in any case, also carries them through the inclement winters of temperate regions of the world. A behavioural adaptation of small blue butterflies *(Tarucus* spp.*)* and others, is the power of continued flight within the shelter of a small bush, even when the wind is raging outside (Buxton, 1923). Hairy caterpillars are sometimes blown along by the wind, rolled up into a ball, while the larvae of *Prionopteryx nebulifera* and other species in America inhabit long tubes of sand attached to the stems of plants. Small moths are especially numerous in the American deserts. The pupal life of many of them is spent out of sight, underground, but a few, the bag-worms, are conspicuous because of peculiar pupal cases they hang on desert plants.

Not only are flies (Diptera) of the family Muscidae (including the common house-fly pests *Musca domestica* and *Fannia canicularis*) all too plentiful in the desert, especially near oases, but representatives of many other families also occur. These include sand-flies (Psychodidae), crane-flies (Tipulidae), hover-flies (Syrphidae), long-snouted Nemestrinidae which probably suck nectar, and bee-flies (Bombyliidae) many of which are black in colour. Robber-flies (Asilidae), Therevidae, Empidae and other predaceous species are not uncommon whilst blood-sucking horse-flies (Tabanidae) constantly attack camels, horses or donkeys. One of the bot-flies (Oestridae) develops in the nose of the camel. The mature larva is sneezed out onto the sand where it pupates. 'Worm-lions' are larvae of Rhagionid flies which live in pits like ant-lions to which they show a striking parallel in their behaviour. The genus *Vermileo* is represented in North Africa and North America, *Lampromyia* in North and South Africa, *Vermitigris* in south-east Asia.

Without doubt, beetles of the family Tenebrionidae are the insects best adapted to desert life. Second to them come dung-beetles and burying-beetles of the families Scarabaeidae, Geotrupidae and Trogidae, while blister-beetles (Meloidae), ground-beetles (Carabidae) and tiger-beetles (Cicindelidae) are also strongly represented. The well-known Egyptian scarab-beetle *(Scarabaeus sacer)* acts as a scavenger by breaking up and burying the droppings of camels, goats and other animals. Many American Scarabaeidae have become secondarily adapted to eating vegetation and fallen leaves. Some genera, such as *Peltotrupes,* may burrow to a depth of 3 metres or more in the soil. Beetles of the family Histeridae are also found on dung or carrion where they prey on the maggots and other insect larvae developing there.

Blister-beetles (Meloidae) are vegetarian. They can secrete a disagreeable, oily fluid from the joints of the limbs. This contains cantharidin which raises painful blisters on the skin. The adults possess conspicuous warning colouration, vivid black and green, brown or blue and red. Cicindelidae and Carabidae are predatory both as larvae and adults. Species of *Cicindela* are skilful fliers, common on sand on the banks of rivers and temporary rainpools, or beside the sea. The larvae inhabit tubes in the sand from which they watch for passing prey. In the American *C. generosa,* the sand grains are cemented with saliva and the burrow serves as a pitfall to trap other insects. Many of the desert Carabidae are wingless. Particularly striking of the Palaearctic fauna are *Anthia venator* and *A. sexmaculata,* gigantic nocturnal predators with a conspicuous black and white warning coloration which is associated with an evil flavour.

The family Tenebrionidae contains the typical black darkling beetles, so common throughout all the deserts of the world (Fig.16). These insects are able to live on dry food without any water. The majority such as the Saharan species of *Blaps, Pimelia grandis* and *Ocnera hispida* are crepuscular or nocturnal (p. 101), but the long-legged *Adesmia antiqua* is day-active, except in extremely hot conditions when it burrows in the sand (Cloudsley-Thompson, 1963b). *Sternocara phalangidium* of the Namib desert has exceptionally long legs while *Stips stali* has short legs and is adapted for burrowing (Koch, 1961). Species of *Eleodes* in America are often active during the hottest parts of the day. Wings are generally absent throughout the Tenebrionidae and the elytra are fused. The sub-elytral air-space not only serves for purposes of insulation but is of even greater importance in reducing the amount of water lost through transpiration, for the spiracles open into it (Ahearn, 1970; Ahearn and Hadley, 1969; Cloudsley-Thompson, 1964b; Hadley 1970) (p. 194). Tenebrionidae are omnivorous, feeding on vegetable matter, carrion and dung. Some species show remarkable morphological adaptations to desert life (Edney, 1971).

Arachnida. Of all animals the scorpion is, perhaps, most symbolic of the desert, even though some species occur in the wet tropics, and others in subtropical regions. Nocturnal in habit, scorpions spend their days in sheltered retreats. They are strictly carnivorous, but do not usually go to seek their food. Instead, they tend to wait for the insects that enter their lairs to hide. When hungry, however, they may emerge at night and walk about with claws extended, ready to grip the prey which is then subdued, if necessary, with the sting. There are two types of scorpion poison. One is local in effect and comparatively harmless to man: the other is neurotoxic and resembles some kinds of snake venom. It also has a haemolytic action, destroying red blood corpuscles, and can be dangerous. The venoms of the Neotropical *Centruroides* spp., of *Androctonus australis* in North Africa, and of *Leiurus quinquestriatus* south of the Sahara, are as toxic as that of a cobra, but the quantity injected is smaller (Cloudsley-Thompson, 1965b).

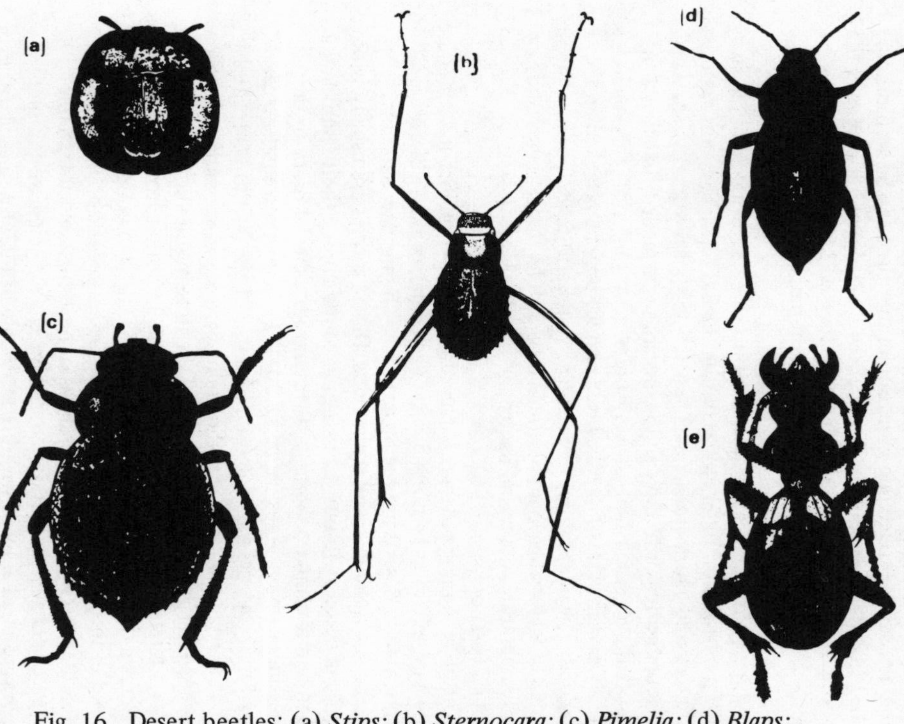

Fig. 16 Desert beetles: (a) *Stips;* (b) *Sternocara;* (c) *Pimelia;* (d) *Blaps;* (e) *Anthia* (From Cloudsley-Thompson, 1969).

Even more than scorpions, Solifugae are typical animals of the desert. Sometimes known as 'jerrymanders', 'false-spiders', 'wind-scorpions', or 'camel-spiders', the long-legged species are familiar to all who have travelled in arid regions. A large *Galeodes arabs* or *G. granti* (Fig.17) whose formidable appearance is enhanced by its unusual hairiness and bulk can, with its limbs, span a width of five or six inches. These creatures avoid fertile oases and seem to prefer utterly neglected places where the soil is broken and bare. Sometimes they can be seen running so fast that they resemble balls of yellow thistledown blowing over the desert. Often, when going at full speed, they will stop abruptly and begin hunting about like a dog checked in mid-course by the scent of game.

As their name indicates, most Solifugae are nocturnal and hide away in deep burrows or under stones during the daytime. They are exclusively predatory and carnivorous, having an extraordinary voracity: they will continue feeding until their abdomens are so distended that they can scarcely move. Almost any insect, spider, scorpion or other 'camel-spider', as well as small lizards, birds or mice may be attacked and devoured. The adaptations of Solifugae to desert life include considerable powers of water-conservation, the ability to tolerate extremes of drought and temperature, nocturnal habits and a wide range of diet (Cloudsley-Thompson, 1961b,c).

Of the remaining orders of Arachnida, only the spiders and mites are important inhabitants of the desert. The spider families chiefly represented in deserts are the Sparassidae, Thomisidae, Salticidae, Lycosidae, Agelenidae and Argiopidae. Desert species are usually white or pale in colour without the markings found among their relatives that live in humid climates. Some of them have brushes of hairs on the undersides of their limbs which facilitate movement on sand. At least one of the larger Sparassidae *(Cerbalus* sp.*)* of the Namib desert digs a tube in the sand and cements the grains with a criss-cross of webbing. Loose sand is pushed up the slope of the tube by the mouthparts of the spider. Others live on the trunks of *Acacia* and other trees.

Crab-spiders (Thomisidae) lie in wait for their insect prey on the ground or in vegetation. They usually possess cryptic coloration and match their background so closely as to be almost invisible. Some species can change colour quite quickly. The Salticidae or jumping-spiders are small animals with broad, square heads, extremely large eyes and short, stout legs. They have very keen sight and stalk their prey from afar whereas wolf-spiders (Lycosidae) hunt in the open and overcome their prey by sheer strength. They have longer bodies and limbs but their eyes are smaller. Agelenidae construct funnel-shaped cobwebs consisting of a triangular sheet with its apex rolled into a tube in which the spider awaits its prey. The orb-webs of the Argiopidae are

Fig. 17 The camel-spider *(Galeodes granti)*.

usually attached to vegetation. That of *Argiope lobata,* a common inhabitant of sand dunes in the Mediterranean region is extremely large. The spider has an irregular shape which, combined with its sandy colour, renders it extremely inconspicuous.

Mites are sometimes found in desert soils, but they are not numerous. Adult giant velvet mites *(Dinothrombium* spp.*)* often appear in the deserts of Africa and America a week or two after rain and probably feed on termites and other insects. The larvae are parasitic on grasshoppers (Tevis and Newell, 1962). The adults dig burrows where the sand is damp. Their scarlet coloration has a warning function and is associated with repugnatorial glands which render them distasteful to enemies such as scorpions and 'camel-spiders' (Cloudsley-Thompson, 1962b). In contrast, many species of tick are well-adapted to life in arid places. When not attached to their hosts, they show remarkable powers of water-conservation and some species can live for ten years or more without food or drink.

Reptiles. Agamid lizards are widely distributed throughout the deserts of the Old World. They are terrestrial in habit, flattened in shape and mostly feed upon insects and spiders. Some species, in particular the scaly-tailed lizards of the genus *Uromastix* may be entirely herbivorous. In these the short, thick tail is covered with whorls of large scales while the body is much flattened and the head smooth and covered with very small scales. Like chuckwallas *(Sauromalus* spp.*)*, scaly-tailed lizards inflate themselves in rock crevices from which it is extremely difficult to extract them. The scales on their tails assist in wedging them securely. True lizards (Lacertidae) are also extremely common throughout the deserts of Asia and Africa. Many of them show marked adaptation to desert life (p. 103).

Iguanid lizards appear to fill a similar ecological niche in America to that occupied by the Agamidae of the Old World. Again, most species are insectivorous, but genera such as *Iguana* and *Basiliscus* appear to be largely herbivorous. The genus *Phrynosoma* includes the 'horned toads' whose flattened bodies are covered with spiky scales. *P. cornutum* of the Sonoran Desert has five spikes on each side of the head while *P. coronatum,* an inhabitant of California, has an additional smaller spine between the two occipitals. The desert iguana *(Dipsosaurus dorsalis)* is abundant in valleys and plains where creosote bushes *(Larrea divaricata)* occur with rodent holes and burrows under them. Adaptations of iguanid lizards of the genus *Uma* include elongate, valvular scales fringing the edges of certain toes thus widening them, serving as sand-shoes and assisting movements beneath the surface of the sand. Nasal valves restrict the entry of sand into the nasal passages which, in turn, are specialised by being convoluted and having absorbing surfaces that reduce loss of moisture through the nostrils. The head is wedge-shaped, which assists in sand swimming, as do the enlarged lateral scales on the legs and tail. These tend to force the animals down into the sand when the limbs are moved. Similar adaptations are found among desert-inhabiting members of other lizard families (Buxton, 1923). In many sand-dwelling lizards, including skinks and geckos, the eyes show a remarkable modification in the form of a window in the lower eyelid (p. 104).

Geckos (family Gekkonidae) are widespread throughout the tropics and several species have become adapted to desert life. The heads of these lizards are broad, their bodies flattened and their toes equipped with pads which adhere by means of friction, thus enabling the animals to run up smooth, vertical surfaces. In the genus *Palmatogecko* the feet are completely webbed for support on loose sand (p. 104). Most geckos are active at dusk and during the night, a behavioural adaptation of especial value in hot, dry climates but *Tarentola mauritanica,* a species common on both sides of the Mediterranean, although usually active in the evening, sometimes basks in the sun and comes into the light to catch insects. Geckos are able to withstand considerable desiccation and starvation. *Tarentola annularis* of the Saharan region may lose over half its body weight without ill effect and then recover when food is again available (Cloudsley-Thompson, 1965c).

Several quite unrelated kinds of desert lizard have adopted a snake-like form of locomotion. The body is covered with smooth scales which cause little friction and the legs may be reduced as in species of *Seps* or even lost, as in the genus *Typhlosaurus* of South-West Africa, so that movement through the sand is accomplished entirely by wriggling the body. The small *Anniella pulchra,* which inhabits the sand dunes of California and Arizona, can move extremely rapidly in this way, but spends much of its time lying in wait for its insect or spider prey with only the fore part of its head exposed. Perhaps the most imposing of desert lizards is the African monitor *(Varanus griseus)* which may reach a length of nearly 2 m. It is a speedy and rapacious creature, and will eat any other animal that it is strong enough to overcome.

Although a certain amount of water is lost from reptiles by evaporation through the skin, perhaps ten times the amount formed by metabolism, it is doubtful if this contributes much to cooling the body. Small reptiles have a higher rate of water-loss per unit body weight than larger ones.

Like lizards, desert tortoises avoid the heat of the sun by burrowing deeply in the ground. The earth is scraped loose with the forefeet, whereupon the animal turns round and pushes it away with its carapace or 'shell'. Advantage is often taken of initial excavations by other animals such as ground-squirrels. During the morning and evening when the air is cool, tortoises warm their bodies by basking in the sunshine. They do not normally drink because they obtain enough moisture from the succulent plants on which they feed. They can, however, ingest large quantities when desiccated. The American desert tortoise *(Gopherus agassizi)* exists without water during the entire dry season. The eggs are laid in early summer and hatch in three or four months. It takes about five years after hatching for a young tortoise to develop a hard shell and about fifteen to twenty years for it to reach maturity.

The Greek tortoise *(Testudo graeca)* occurs throughout the Mediterranean region and Asia Minor, its range extending eastwards as far as Iran. Although normally feeding on juicy plants, these tortoises have been found eating the astringent green fruits of the dwarf palm *(Chamaerops humilis)*. They spend much of the time basking in the sun but seek the shade when their temperature begins to approach lethal limits. At all times they rise late and retire early, being absolutely diurnal in their habits. Evaporative water-loss from the African tortoise *(Testudo sulcata)* increases greatly when the air temperature exceeds 40.5°C (105°F) because the temperature of the body is maintained at this level by copious salivation which wets the head, neck and front legs. In this species the eggs are laid during the autumn and hatch the following summer at the time of the annual rains (Cloudsley-Thompson, 1970; 1971a).

Snakes are much less common than lizards. Nevertheless they comprise an important element of the desert fauna. They are the most highly specialised of the carnivorous reptiles and are exclusively meat-eaters. Although their tolerance of high temperatures in general is lower than that of lizards or tortoises – for example the lethal temperature of the desert rattlesnake *(Crotalus cerastes)* is 41.5°C while that of lizards inhabiting the same area is 45-47.5°C – snakes are even more adept at insinuating themselves into holes and crevices. Sidewinding locomotion is characteristic of *C. cerastes* and horned vipers of the genus *Cerastes* in the Great Palaearctic Desert. It produces a high speed, halves the area of body in contact with the hot soil and enables the prey to be approached unobtrusively (Cowles, 1957). Like lizards, snakes possess a relatively impervious skin which is especially impermeable in desert species, and little water is used in excreting urinary waste matter. Their carnivorous diet is rich in moisture, and evaporation is greatly reduced by daily or seasonal quiescence in a relatively cool and humid burrow (see discussion in Cloudsley-Thompson, 1971a).

The two most important groups of non-poisonous snakes found in deserts are the worm-snakes (Typhlopidae and Leptotyphlopidae) and the harmless members of the Colubridae. These snakes, having no special method of killing their prey, eat it alive so that it ultimately dies of suffocation or from the action of digestive juices. Worm-snakes have a tropical distribution; some of them, such as *Leptotyphlops* spp., extending into the desert where they lead secretive lives under logs and stones or burrowing into the ground. Their eyes are rudimentary, their dentition reduced and they feed on termites and other small insects.

Desert Colubridae often show some of the adaptations to burrowing in sand that are found among lizards. They are usually more slender than their relatives from more humid places. They are active creatures belonging to a vast assortment of species. In the related back-fanged snakes, the prey has to be gripped before the poison, which usually has a paralysing effect, can be chewed into the wounds made by the teeth.*

Birds. Compared with most groups of animals, birds show little specialisation in form or appearance to desert life. Apart from being generally paler in colour, most species are hard to distinguish from their relatives in humid climates. The most numerous are insect-eating species followed by seed-eaters and, lastly, least numerous are the carnivores. Many of the latter feed chiefly upon reptiles. Golden eagles *(Aquila chrysaetos)* and lanner falcons *(Falco biarmicus),* for example, live in the Sahara to a large extent upon spiny-tailed lizards *(Uromastix* spp.*).* The Sonoran white-rumped shrike *(Lanius ludovicianus)* impales its lizard prey on sharp thorns, while the road runner *(Geococcyx californianus),* a relative of the cuckoos, like the African secretary bird *(Sagittarius serpentarius)* regularly feeds on snakes.

Although many species of birds have been recorded from desert regions, their distribution there is usually closely related to the presence of surface water for drinking. Where there is no such water within their range of flight, birds are very scarce in contrast to the relative abundance of reptiles and small mammals. Water loss by evaporation is the most serious physiological factor limiting their distribution in arid regions. Most birds are active in the daytime and, because they do not have burrowing habits, they cannot escape from the mid-day heat as do small mammals and reptiles. Even though they rest in the shade as much as

---

*It is not always realised that most snakes are non-poisonous. In fact, only three groups are venomous – the back-fanged Boiginae, the deadly elapid snakes (such as the kraits, cobras and mambas), and the vipers and rattlesnakes in which fang development has attained its greatest perfection. It is questionable whether elapid or viperid poison is the more lethal, for the physiological effects of the two are quite different. The former is neurotoxic and induces paralysis; the latter causes collapse and heart failure, depending upon the amount injected. Snake poison contains a large number of toxic compounds and varies from one species to another. Viperid poison usually contains an ingredient that causes clotting of the blood while elaphid venom more often contains an anti-coagulant and, at the same time, causes haemolysis or breakdown of the red blood corpuscles.

possible during the hottest part of the day, they are exposed to much more radiant energy than are burrowing rodents. (Owls and nightjars are an exception and secrete themselves in rock clefts and fissures during the daytime.)

Early in the morning, most desert birds may be seen feeding actively but, later in the day, the smaller species shelter in trees with hanging wings and open beaks, while large ones, such as eagles, hawks and vultures, circle high in the sky where the air temperature is considerably lower than it is near the ground. Carnivorous and insectivorous birds obtain plenty of water with their food: it is the seed-eaters that are faced with the greatest problem of survival.

Owing to its large size, the ostrich *(Struthio camelus)* (Fig. 18) cannot obtain shelter from the heat of the day in the manner of smaller birds. Nor is it so mobile. Although it must drink or eat very succulent food, nasal salt-excreting glands enable it to live off brackish or even salty water. Like the camel, it can tolerate considerable desiccation and rapidly make good the loss in weight suffered during dehydration, (Cloudsley-Thompson and Mohamed, 1967). Cooling is achieved by panting: there is a counter-current flow of air and blood in the lung, and a functional shunt system prevents alkalosis (decrease in carbon dioxide of the blood) (Schmidt-Nielsen *et al.*, 1969). The ostrich, a native of Africa and the Middle East, is not entirely a desert bird for it is found over Africa generally wherever the country is open and dry. Behavioural reflexes, including erection of feathers and wing drooping, are important in enhancing its physiological adaptations to heat (Luow, Belonje and Coetzee, 1969).

Mammals. Small mammals are able to avoid the midday heat of the desert by burrowing. Their physiological problems are therefore concerned with water shortage rather than with temperature stress. Many of them can exist without drinking at all, the most remarkable in this respect being jerboas and kangaroo-rats.

There are large numbers of rodents inhabiting different deserts of the world which have characters in common, although some of them are but distantly related. Except for the spring-hares *(Pedetes* spp.*)* of East and South Africa, they tend to be about the size of a small rat, with short fore legs and long hind legs adapted for jumping. The number of toes is reduced, the tail long with a terminal tuft of hairs and their gait is characteristically bipedal. The short fore limbs are used for burrowing, the hind legs for jumping while the body is balanced by the tail.

Many desert mammals, including rodents, bats, hedgehogs, foxes, gazelles and addax antelope, possess greatly enlarged tympanic bullae. The significance of this common feature is not entirely clear, but it probably adds greatly to the sensitivity of the ear, especially to sounds of low frequency made by enemies such as owls and snakes. It may also

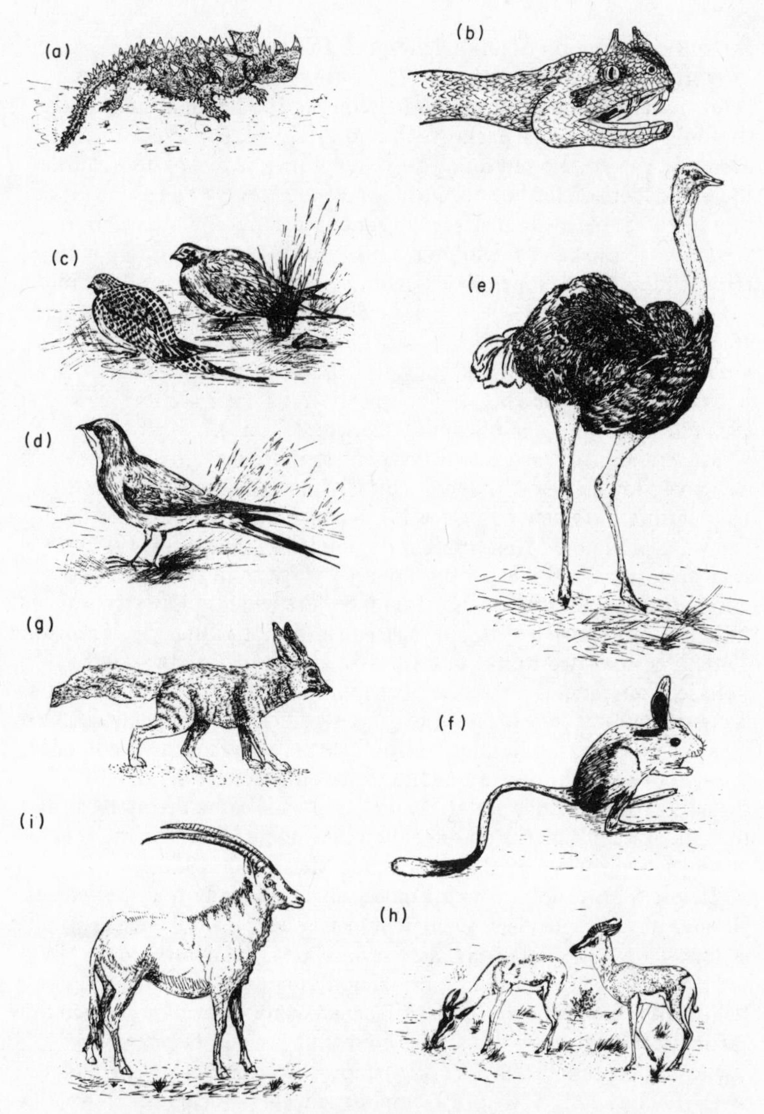

Fig. 18 Vertebrate animals of the desert: (a) *Moloch horridus* (15cm); (b) Head of *Cerastes cerastes* (length of whole snake 30cm); (c) Sand-grouse, *(Pterocles)* (30cm); (d) *Glareola pratincola* (25cm); (e) Ostrich, *(Struthio camelus)* (240cm); (f) Jerboa *(Jaculus)* (15cm); (g) *Fennecus zerda* (30cm); (h) *Gazella dorcas* (50cm); (i) *Oryx leucoryx* (100cm). (Drawings not to scale) From cloudsley-Thompson, 1965.)

aid in the perception of ground vibrations (Prakash, 1959).

North American kangaroo-rats *(Dipodomys* spp.*)* and other small desert rodents can survive indefinitely on dry food without any water to drink. Indeed, on an exclusive diet of dry grain, *D. merriamii* can even gain weight. No physiological water storage is drawn upon, nor is there any increase in the concentration of the blood while living on a dry diet. This means that water is not conserved by the retention of waste metabolites but the urine is almost twice as concentrated with respect to salts as is the urine excreted by the white rat, and 1.6 times as concentrated with respect to urea. *Dipodomys* is even able to utilise sea-water for drinking, as it can excrete such large amounts of salt and yet maintain normal water balance. It can also eliminate an excess load of urea, about 23 per cent, which is nearly four times as much as in man (Schmidt-Nielsen, K. and Schmidt-Nielsen, B. 1950).

The amount of water lost by evaporation through the lungs is extremely low in desert rodents. This is achieved by reduction in the temperature of the expired air as it leaves the tip of the nose, so that less water is required to saturate it (Schmidt-Nielsen, 1964; 1972). Kangaroo-rats are nocturnal and do not emerge from their burrows during the day. Measurements of temperature and humidity recorded in their burrows showed that, although the air is not saturated, its moisture content is from two to five times as high as that of the atmosphere outside. Consequently, the rate of evaporation of water from the lungs is considerably reduced. This is of great significance. If the animals were breathing the air outside their burrows, with its low moisture content, the rate of evaporation from their lungs would exceed the rate of formation of metabolic water. As long as they breathe the moist air in their burrows during the daytime, however, there is an ultimate gain in water (Schmidt-Nielsen, B. and Schmidt-Nielsen, K., 1950).

Desert rodents do not sweat: indeed, it seems likely that the general absence of sweat glands in small mammals results from the necessity that is imposed by their relatively large surface area, to conserve water. In order to maintain a constant, normal body temperature when the air temperature is around 40°C (104°F), a kangaroo-rat would have to lose 20 per cent of its body weight per hour. But there is an emergency regulatory process for, if the body temperature approaches the lethal level of about 42°C (107.5°F), copious salivation occurs which wets the fur of the chin and throat. The cooling effect of this may keep experimental animals alive for up to half an hour at temperatures fatal to other small rodents: but, of course, because of the rapid heat gain of small animals and the limited amount of water available, the time limit on the mechanism is severe. Possibly an animal driven from its burrow by a predator may have a better chance of survival with such a mechanism than without. Clearly, water can be used for heat regulation only in the greatest emergency and evaporation normally takes place only through the lungs (Schmidt-Nielsen, K. and Schmidt-Nielsen, B., 1952).

The adaptations of the Egyptian jerboas *(Dipus aegyptus* and *Jaculus jaculus)* (Fig. 18) to their desert environment are similar to those of the American kangaroo-rats. Underground living and nocturnality are partial solutions that jerboas have found to the three main problems of the desert – scarcity of water, rarity of food and solar radiation. The jerboa can live indefinitely on a dry diet, barley and wheat grains (containing 11-12 per cent water) by eating little and excreting little. The urine is concentrated and the faeces dry (Kirmiz, 1962).

Little is known about desert carnivores, but there is no doubt that they obtain considerable quantities of water from the body fluids of their prey. In addition, they tend to have a more varied diet than their relatives from more temperate regions. Carnivores include foxes, jackals, hyaenas, coyotes, small cats, badgers, skunks, ferrets, some carnivorous marsupials and the Australian dingo. Of these, only foxes are found in extremely arid regions and may be entirely independent of drinking water. The delightful fennec fox *(Fennecus zerda)* (Fig. 18) of the Sahara exhibits a number of adaptive characters which are paralleled by the American kit-foxes *(Vulpes velox* and *V. macrotis)* (p. 205). The fennec is much smaller than its relatives from temperate climates, has well-developed sense organs, large eyes and ears, and spends the day in a deep burrow, thus avoiding extremes of heat. Like other desert species, it is a pale, sandy colour. It is more catholic, too, than other foxes and lives mainly on insects, lizards, rodents, dates and so on. Its liking for sweet things provides an explanation of the fable of the fox and the grapes. The young are born in burrows at the time of the spring rains: the adults weigh less than one kilogramme. If they get over-heated, fennecs pant like dogs. They excrete a highly concentrated urine, and the same is probably true of kit-foxes (Schmidt-Nielsen, 1964).

A number of rodents and other small mammals are able to live in the desert by subsisting on moist food. These include the sand-rat *(Psammomys obesus)* of North Africa which lives and nests in places where the vegetation consists of succulent plants. These are usually extremely salty, but sand-rats eat them in great quantities and secrete a copious urine which may be up to four times the concentration of sea-water. The American pack-rats *(Neotoma* spp.*)* and ground-squirrels *(Citellus* spp.*)*, which feed on juicy cholla fruits, do not possess the same ability to eliminate large quantities of salt but, nevertheless, excrete a concentrated urine. Pack-rats protect the entrances to their burrows by piling up stones and pieces of prickly cactus. Ground-squirrels aestivate in their burrows during the desert summer.

Other small American desert rodents are grasshopper mice

(Onychomys spp.*)*, so called on account of their insectivorous diet which provides them with all their water requirements. The desert hedgehog *(Hemiechinus aruitus)* of North Africa also obtains water from its food, while the crest-tailed marsupial mouse or mulgara *(Dasycercus cristicauda)*, which inhabits the most arid central parts of Australia, lives predominantly on insects supplemented by occasional lizards and small rodents. Like the grasshopper mouse, it is able to excrete in a relatively small volume of water the large amounts of urea, which result from its carnivorous diet. In common with most other deserticolous mammals, it is of a pale sandy colour (Schmidt-Nielsen, 1964).

Although many species of desert hares and rabbits excavate tunnels and burrows, the American jack rabbits *(Lepus californicus* and *L. alleni)* remain above ground and have no underground retreat. They live in areas where no free water is available and depend upon the moisture obtained with their green food. It is not entirely clear how they can survive in the desert, but the suggestion has been made that their very large ears with a network of blood vessels may serve to radiate heat to the sky while the animals are resting in the shade. Large ears are a characteristic of many desert animals and the Saharan hare *(Lepus capensis)* has ears much larger than those of its relatives of temperate climates.

The larger mammals of the deserts of the Old World include antelopes, gazelles and wild asses. In the arid regions of North America their ecological niche is occupied by the pronghorn *(Antilocapra americana)*, formerly very common in the prairies but now extremely rare, and the mule-deer *(Odocoileus hemionus)*. The larger herbivores of the Australian deserts are marsupial kangaroos and wallabies, whose young are born at an early stage of embryonic development and carried by the mother in a pouch or marsupium until they are big enough to fend for themselves.

None of these animals is able to escape the rigours of the desert climate by burrowing, but they have low water requirements, and their mobility enables them to travel long distances to obtain drinking water if required. The dorcas gazelle *(Gazella dorcas)* (Fig. 18), one of the smallest of the gazelles, with a shoulder height of up to 60 cm (24 in.), is widely distributed throughout the Sahara Desert. The back and sides are sandy reddish-brown, the belly white, the two colours being separated by an indistinct dark stripe. The horns of the male are stouter and more curved than those of the female. Although in Morocco, Palestine and other desert regions of the Mediterranean basin, dorcas gazelles do not require water and can obtain sufficient moisture for their needs from succulent roots and plant material, in the Sudan they lose weight steadily on dry food when deprived of water. After five days' desiccation, a maximum of 1.5 litres of fresh water can be ingested and smaller quantities of saline water are taken. With increasing dehydration, body temperature tends to fluctuate and there is some degree of hyperthermia or increased temperature, the urine becomes concentrated, faecal pellets smaller and drier, and food intake is reduced. Feeding ceases when 14-17 per cent of normal body weight has been lost and the animals appear weak and emaciated. This may take up to twelve days in winter but, in summer, gazelles cannot survive for more than five days without drinking (Ghobrial, 1970). It seems that, whereas camels (p. 97) are able to survive on a low water intake, the physiological adaptations of the dorcas gazelle are less well marked and their survival depends upon speed and mobility so that, when necessary, they can travel great distances to water. Gazelles can run at 50 km. p.h. They migrate from the western Sudan to the Nile during the dry season. During the rains herds of two dozen or more can be seen but in dry weather they are more solitary. In very hot weather they have been seen to cool themselves in the Red Sea, but they do not drink salt water. A dwarf race of the Arabian gazelle *(Gazella arabica)*, one third of the normal weight, is said to inhabit islands in the Red Sea where no fresh water is available for drinking.

The odd-looking Saiga antelope *(Saiga tatarica)* (Fig.19), which roams the arid plains of western Asia, is about the size of a fallow deer. Its coat is dirty yellowish in summer, longer, thicker and paler in winter. The most peculiar feature is a greatly elongated and swollen nose with very wide nostrils separated and turned back so that sand is excluded while the animal is grazing. In common with other animals of steppe and desert where concealment is scarce and sources of water infrequent, it has great swiftness and can cover long distances.

Among the more interesting of the ungulates of the African and Asian deserts and steppes are the wild asses. Several species are known, including the onager *(Equus onager)* of which different subspecies occur from Central Asia to North-West India, Baluchistan, Iran and Syria. This is the "wild ass" of the Bible: it is white with a large yellowish area on each flank and a black dorsal stripe, mane and tail. It is gregarious, the herd being led by an old stallion. The kiang *(Equus kiang)* is a deep reddish-brown in colour and is more solitary. It inhabits the high desert plateaux of Tibet, Ladak and Sikkim. The kulan or chigetai *(Equus hemionus)* is smaller. Its range extends from the steppes of Transcaspia, Transbaikal and Mongolia into the Gobi desert. African wild asses are represented by two subspecies of *Equus asinus,* the Nubian and Somali wild asses. It seems probable that none of the wild asses of today are pure descendants of the original, true wild ass, and there has been much interbreeding with domestic asses (Matthews and Carrington, 1970).

Asses are sure-footed, long-eared animals that resemble the camel in the ability to tolerate a considerable degree of dehydration and to

Fig. 19 Saiga antelope *(Saiga tatarica)*.

withstand a water-loss of 30 per cent of the body weight. Their drinking capacity also is impressive and, within a few minutes, they can ingest more than a quarter of their body weight. Asses appear to lose water more rapidly than camels, however, because the fluctuations in their body temperature are smaller. Their fur coats are thinner and provide less effective insulation and their behavioural adaptations which reduce heat gain from the environment are less extreme (Schmidt-Nielsen, 1964).

The Australian red kangaroo *(Macropus rufus)* and other species range widely throughout the inland desert regions of the continent. They can exist on the water content of their vegetarian diet, supplemented by infrequent drinking. Their bipedal, leaping gait, in which the body is carried well forward and counterbalanced by the massive tail, enables them to travel long distances at 30 km.p.h. In short bursts a speed of 50 km.p.h. can be achieved with leaps of over 7 m (25 ft).

The quokka *(Setonix brachyurus)* is a medium-sized marsupial, about the size of a rabbit, which inhabits the offshore islands and coastal regions of south-west Australia where no fresh water is available and moisture obtained from the vegetation is supplemented by sea water. In hot weather, the body is cooled by sweating, supplemented by copious salivation, licking of the feet, tail and belly, a trait also found amongst kangaroos (Bartholomew and Dawson, 1956).

Camels are large, even-toed ungulate mammals with humped backs and digitigrade feet in which there are only two toes, the third and fourth. These are united by thick, fleshy pads which prevent them from sinking in soft sand and are tipped with nail-like hooeves. Camels were first domesticated by man in pre-historic times. Two species are recognised: (a) The Arabian camel or dromedary *(Camelus dromedarius)* which is widespread throughout the Middle East, India and North Africa and possesses a single hump. (b) The bactrian camel *(Camelus bactrianus).* This is a heavily-built, two-humped animal which inhabits the deserts of central Asia where the winters are very cold. It has a longer, darker winter coat, shorter legs and seldom measures more than 2.1 m (7 ft.) from the ground to the top of the humps. This is about the height of the shoulder in the taller and more slender dromedary.

The dromedary is known only in domestication, but the small humps and feet and short, brown hair of the bactrian camels of the Gobi desert indicate that they are genuinely wild and not merely feral – that is, descendants of domesticated stock that have escaped over the centuries.

When moving fast, camels pace. Like the giraffe and brown bear, they raise both legs on the same side of the body and advance them simultaneously while the weight is supported by the legs of the opposite side. In this way a speed of up to about 8 km.p.h. may be achieved; but it cannot be maintained for more than a few hours.

Although they chew the cud, camels differ from true ruminants in that they lack an omasum or third section to their stomachs. The smooth-walled rumen or anterior section has small sacs or diverticula leading from it. These were formerly called "water sacs" because of an erroneous hypothesis, dating from Pliny's *Historia Naturalis,* that they served the function of water storage. These glandular rumen sacs contain a fluid having the same salt content as the rest of the body. It is like green pea soup in appearance and is quite repulsive. To the desert traveller who has no water, however, any fluid is attractive. The many tales of people who have saved their lives by killing their camels to drink the fluid in the rumen sac may therefore well be true (Schmidt-Nielsen, K. *et al.*, 1956).

Equally erroneous is the idea that the camel stores water in its hump, or that the fat from which the hump is composed is essentially a water store itself. On oxidation, this fat produces metabolic water, but the extra oxygen that must be used in the process involves, in turn, an extra loss of water through the lungs which just about cancels any gain from oxidation of the fat. The camel's hump is actually a food store which, by being concentrated in one large depot and not distributed as a subcutaneous layer of fat, allows the rest of the body to act as a radiator for cooling purposes (Schmidt-Nielsen, 1962).

The rate of urine flow is low in camels and little water is lost with the faeces. Instead of eliminating all the urea produced in metabolism the camel, like ruminants, can utilise it for microbial synthesis of protein. In this way the amount of water excreted is reduced and greater use made of the food.

The coarse hair on the camel's back acts as a barrier to solar radiation and slows the conduction of heat from the environment. It is well ventilated so that evaporation of sweat occurs on the skin where it provides maximum cooling. At the same time the camel avoids undue water-loss by allowing its temperature to vary over a range greater than that of other mammals and sweating does not begin until the body temperature has risen to 40.7°C (105.3°F). Thus heat is stored during the day and lost at night when the environmental temperature is lower. Moreover, when the camel's temperature rises, the difference between it and that of the air is reduced so that less sweat is required to prevent a further increase in body temperature.

The camel can tolerate a much greater depletion in body water than most other mammals and may, without ill effects, lose about 30 per cent of its body weight (100 kg out of 450 kg) as compared with about 12 per cent in man. It also has an unusual drinking capacity and can assimilate 114 l. (25 gallons) or more in a very short space of time. The blood and tissue fluids become rapidly diluted to an extent that would cause other mammals to die from water intoxication. This ability is related to the physiology of the blood, whose red corpuscles or erythrocytes are unusually resistant to dilution and can swell to twice their initial volume without rupturing.

With most mammals subjected to high temperatures in dry air, desiccation proceeds steadily while the body temperature remains constant. As water is lost through evaporative cooling, however, the blood gradually becomes more viscous until it cannot circulate quickly enough to carry away metabolic heat to the skin. At this point the temperature suddenly rises and 'explosive heat death' results. In camels, this is avoided by a physiological mechanism which ensures that water is lost from the tissues only, while the blood volume remains fairly constant (Schmidt-Nielsen, B. *et al.*, 1956).

Thus the adaptations of the camel to its desert environment do not involve independence of drinking water but the ability to economise the water available and to tolerate wide variations in body temperature and water content. In winter, when the temperature is comparatively low and water is not needed for heat regulation, camels become independent of drinking water for several months. In summer, the length of time between drinks depends on the environmental temperature and the amount of work that the animal is called on to perform.

Seasonal rhythms

Seasonal changes of the environment in hot, arid regions may result from one of two factors: seasonal precipitation and a consequent outburst of plant growth, or a seasonal period of cooler, less severe weather, even if rain does not fall (Cloudsley-Thompson, 1969).

These changes are reflected in the numbers and stages of development of the fauna so that populations of adults reach their peak at the time of the rains. For example, adults of tenebrionid beetles *(Adesmia bicarinata)* begin to appear in small numbers in Egypt during late October. Their population then gradually increases, reaching a peak in March. By the end of May the beetles have disappeared completely and only dead bodies can be found. During the hot season, the life-cycle is continued by the larval and pupal stages of development (Hafez and Makky, 1959).

In the case of life-cycles such as this, which are extremely common among desert insects and arachnids, diapause and quiescence may or may not occur in the developmental stages. Either way, a vernal rain fauna appears at the time of inflorescence when the desert is transformed by an abundance of plant and animal life. Flowers are visited by Lepidoptera, bees, wasps, hover-flies (Syrphidae), bee-flies (Bombyliidae) and other Diptera. The droppings of camels and goats are rolled away by dung beetles and grass seeds harvested by greedy ants. Termites extend their subterranean galleries to the soil surface and indulge in nuptial flights, while predators such as scorpions, Solifugae, spiders, ant-lions (Myrmeleonidae), bugs, wasps, robber-flies (Asilidae) and predatory beetles glut themselves on an abundance of food. With the rains, too, come swarms of locusts *(Schistocerca gregaria)* which breed in the damp sand. The ephemeral vegetation is devoured by hordes of caterpillars and crickets, and the air buzzes with an abundance of flies, wasps and bees rarely seen at other times of year. Moths and butterflies are plentiful, migratory birds appear and build their nests, and most of the resident reptiles, birds and mammals produce their young while the harshness of the desert is briefly alleviated.

Many species of birds inhabiting desert of semi-arid areas have adapted their reproductive physiology to take advantage of the unpredictable and sporadic rainfall that may occur at irregular intervals. Breeding may be completely suppressed in rainless years and then take place, two or three times in quick succession, after a sudden downpour (Lofts and Murton, 1968). In the desert areas of Arizona, Albert's towhee *(Pipilo alberti)* has been seen to nest within ten to fourteen days after heavy rain in March and April, and continue breeding in a scattered fashion until late summer if rainy conditions prevail.

A similar breeding response to sporadic rainfall has been noted among xerophilous species of Australia so that 30 out of 60 species examined attained full breeding condition within a month of rain falling (Keast and Marshall, 1954). In another area where only a small quantity of rain (insufficient for full breeding) had fallen, specimens collected showed a diversity of abnormal responses. In contrast, the sooty falcon *(Falco concolor)* of North Africa, can breed under the harshest conditions.

The eggs of birds, particularly of smaller species, are vulnerable to overheating by exposure to direct sunlight. Only the largest, those of the ostrich, can withstand prolonged exposure to the desert sun. Various devices are therefore adopted to shelter the eggs. Larks nest as far as possible under shrubs and bushes, wheatears in holes or small caves. Certain species in Arizona and California are dependent upon holes in the saguaro cactus *(Carnegiea giganteus).* These holes are hollowed out by two species of woodpecker, which are thereby able to colonise desert areas where no other large plants grow. Various other species make use of deserted woodpecker holes. These include the elf owl *(Micropallas whitneyi),* the screech owl *(Otus asio),* a sparrow-hawk *(Falco sparverius)* and a flycatcher. Of these, only the elf owl is found exclusively in deserts in which the giant cactus grows, but the others all use old woodpecker holes as nesting sites where these are available.

Of the birds which breed in the open without protection, most species sit on the nest from the time the first egg is laid, so that it is never exposed to day-time heat or nocturnal cold. The pratincole

(Glareola pratincola) and the tern *(Sterna saundersi)* are said to stand over their eggs, thus shading them from the sun.

Sandgrouse *(Pterocles* spp.*)* are common inhabitants of the Great Palaearctic Desert. They nest far away from rivers and lakes, and have adopted an extraordinary method of watering their young. The male rubs his breast on the ground so that his feathers are awry and easily saturated while he is drinking. He then flies back to the nest where the young pass the wet feathers through their beaks and keep changing places until the supply of water is exhausted. Until they can fly, they take water in no other way (Cade and Maclean, 1967).

Although the timing of the reproductive cycle in birds depends primarily upon day-length of photoperiod, breeding among desert species is frequently engendered in response to rainfall or the visual stimulus provided by green vegetation. It has been shown experimentally, however, that xerophilous birds can still be affected by variable photoperiods although under natural conditions other stimuli are stronger and over-ride the influence of light. These opportunist species have also evolved the ability to respond quickly to suitable breeding conditions and maturation of the gonads in juveniles is much more rapid than in most temperate species (Immelmann, 1963). This is not always the case, however, for the double-banded courser *(Rhinoptilus africanua)* breeds continuously in South Africa regardless of weather or season, (Maclean, 1967). For a full discussion of this subject, see the excellent review by Lofts and Murton (1968).

When the mammals come under scrutiny a not dissimilar situation manifests itself. In the matter of reproduction, desert mammals show some interesting adaptations, whose function is to ensure that breeding and the production of young occur at the right season. In Israel, both the rutting of camels and the time of birth, which takes place after a gestation period of twelve months, coincide with the short flush of green vegetation that appears from January to March each year. In most mammals the breeding rhythm is largely exogenous so that, when transported from the northern to the southern hemisphere, they soon adapt their breeding cycles to the reversed conditions (Edney, 1965). In camels, however, the rhythm appears to be more firmly fixed. Bodenheimer (1954) points out that, when camels from an area of winter rains, such as Syria, are transported to the northern Sudan, where there is summer rainfall, they cease almost entirely to reproduce. Certain North African gazelles behave similarly.

It is difficult to see the advantage of such rigid rhythms, and they are not characteristic of all desert mammals. In a good year, for example, some gerbils *(Meriones* and *Acomys* spp.*)* are capable of continuous breeding although reproduction is usually confined to the short spring and seems to be triggered by abundant fresh food.

Seasonal reproductive rhythms are upset by irregular rainfall or drought. In the semi-desert of northern Kenya where rainfall is very unreliable, the naked mole rat *(Heterocephalus glaber)* lives in colonies which greatly extend their burrows during the rains. Food is thus harvested collectively and stored at social nesting sites. Breeding follows the stockpiling of food upon which the females and young are very dependent. It is probable that no breeding takes place in very dry years (Jarvis, 1969).

An interesting method of synchronization with environmental conditions is seen in the desert locust *(Schistocerca gregaria).* Although the interval between fledging and oviposition may be as short as three weeks in these insects, it can be extended to nine months. In Somalia, for instance, delays of three to five months are quite common. Yet, even after such an interval, egg-laying may occur more or less simultaneously at sites hundreds of km. apart. The obvious environmental factors with which this can be correlated is the onset of the rains, but maturation of the gonads begins before the rains actually begin. The signal to which the locusts respond has been shown experimentally to be contact with the terpenoids of aromatic shrubs such as *Commiphora myrrhae.* These terpenoids are in highest concentration just at bud-burst, and it appears that desert locusts respond to their scents, which act as a proximate factor (p. 38) (Carlisle, Ellis and Betts, 1965).

The desert jerboa *(Jaculus jaculus)* breeds just after the rains (October-November) in the Sudan (Happold, 1967). The same is true of other desert rodents of Africa (Happold, 1966) and India (Prakash, 1960). In contrast, *Gerbillus pyramidum* breeds at most times of the year in the Sudan except, perhaps, during the hottest and driest period (March-May), having a much longer breeding season than *J. jaculus,* and perhaps two generations or more (Happold, 1968).

Diurnal rhythms

The survival of small animals in desert regions depends very much upon the section of suitable environments. Most species are nocturnal in habit and pass the day in holes, burrows and other sheltered retreats. Consequently, the fauna of the desert at night is quite different from that during the day. In the desert, where the climatic differences between day and night are more marked than in any other environment, such daily rhythms are especially conspicuous.

Most desert Arachnida and insects confine their activities to the hours of darkness when the temperature is low and the relative humidity of the air comparatively high. This generalisation applies even to species with comparatively impervious integuments, whose relatives in temperate climates are normally active during the daytime.

Scorpions and Solifugae are markedly nocturnal although they can withstand remarkably high temperatures and desiccation. It is difficult therefore to ascribe any function, other than the avoidance of vertebrate animals, to their strict rhythms of activity. Of course, it might be argued that scorpions are protected by their poisonous stings. Poison is not always an effective deterrent to large and powerful enemies, however, which might well trample a scorpion underfoot just as deer will stamp on a snake. And it is well known that baboons and other monkeys become adept at catching scorpions without themselves being stung (Cloudsley-Thompson, 1960b). On the other hand, certain desert beetles, such as *Adesmia antiqua* in the Sudan, can often be seen wandering around in broad daylight. These are distasteful to predators, however, and have extremely hard integuments. Their black coloration has an aposematic or warning function as well as possibly protecting them from ultra-violet light and they are not compelled to be nocturnal in order to avoid predation. These are exceptional, however, and related species such as *Pimelia grandis* and *Ocnera hispida* are strictly nocturnal (Cloudsley-Thompson, 1963b). Animals with moist skins, such as worms and slugs, or with porous outer coverings such as woodlice, centipedes, millipedes, certain mites, soft-bodied insects and amphibians, are usually only active at night, whatever part of the world they inhabit (Cloudsley-Thompson, 1960b; 1962a).

Day-active insects tend to leave the sand when its temperature reaches about 50°C (122°F). Some climb grasses, some dive into holes, whilst others fly about above the ground, making hurried landings to enter their burrows. When confined and prevented from buying themselves or flying away they very soon die (Chapman *et al.*, 1926). There are only a few animals, including some grasshoppers, beetles and spiders that are active during the hottest part of the day in summer. Many of these have long legs which raise their bodies above the scorching sand. The same applies even to some nocturnal and crepuscular or twilight-active animals such as the desert woodlouse *(Hemilepistus reaumuri)*, the large white sparassid dune spider *(Leuchorchestris* sp.*)* and the tenebrionid beetle *Sternocara phalangium* of the Namib desert of South-West Africa (Lawrence, 1959).

The behavioural adaptations and time of activity of desert animals are probably often more important than any physiological specialisations. Indeed, it would be hard to point to any important physiological adaptation to desert life among invertebrates that is not found in at least some non-desert forms.

Most skinks and iguanid lizards are day active and geckoes nocturnal. The scaly-tailed lizard *(Uromastix acanthinurus)* is active during the late part of the night and at dawn in the Tibesti mountains but the skink *Chalcides ocellatus* and gecko *Ptyodactylus hasselquisti* are diurnal in habit although the former shelters from the midday heat. Microclimatic measurements in the holes and retreats of these animals shed light on such differences in behaviour. *U. acanthinurus* burrows deeply in the sand so that the temperature in its hole remains constant at about 35°C (95°F) in June. *C. ocellatus* shelters under fallen leaves at midday, whilst *P. hasselquisti,* the most active of the three species, avoids high temperatures by resting in the shade of rocks. Desert snakes tend to be nocturnal whilst tortoises are day-active. Nevertheless they shelter themselves during the hottest part of the day (Tercafs, 1962).

As we have seen (p. 88), small mammals such as kangaroo-rats, jerboas, gerbils and other rodents, fennec and kit foxes only survive in the desert because they are nocturnal. Birds, except for owls and nightjars, and large mammals such as camels, antelopes and wild asses which cannot burrow, are day active although, again, as far as possible they seek the shade when the sun beats down most fiercely.

Animal adaptations

The problems confronting desert animals are concerned with the necessity to breathe air, to conserve water and, at the same time, to avoid, tolerate or control extremes of temperature. Like plants, many desert animals evade the adverse conditions of the desert summer by aestivating in a state of suspended animation. This dormant state or *diapause* is characterised by temporary failure of growth and reproduction, by reduced metabolism and enhanced resistance to heat, drought and other climatic conditions. During aestivation the mouth of the shell of desert snails is closed by a thick diaphragm which reduces water-loss by evaporation. Desert snails have been known to remain in this state for over five years. The fauna of desert rainpools is dominated by filter-feeding phyllopods, including species of *Triops (= Apus)* whose life span is compressed into a couple of weeks during which the eggs hatch, the animals grow to maturity and lay more eggs before the pools dry up (Rzóska, 1961). These eggs remain in diapause until rain falls again. Most desert insects and Arachnida show diapause in one or other of their developmental stages, so that their period of maximum activity coincides with the time of rainfall which, though erratic, tends to be seasonal.

Although torpor has been described in only one bird, the poor-will *(Phalaenoptilus nuttallii)* (Jaeger, 1965), it occurs in a number of desert rodents such as ground-squirrels *(Citellus* spp.*)* which aestivate during the summer of early autumn. Their body temperature then drops to that of the ambient air, respiration and other physiological processes are greatly reduced along with corresponding water-loss.

Small animals have a very much larger surface area in proportion to their mass than larger ones. Consequently, they cannot afford to use water for evaporative cooling. Arthropods, insects, spiders, scorpions and their allies have relatively impervious integuments, as do reptiles, while birds and small mammals lack sweat glands. They can therefore survive in deserts only by sheltering from the midday heat in bushes, cracks in rocks and holes in the ground. The exploitation of burrowing habits is the most important behavioural adaptation of small animals to desert conditions. This is particularly marked in animals such as the desert centipedes *(Orya* and *Scolopendra* spp.*)* and the desert woodlouse *(Hemilepistus reaumuri)* of North Africa and the Middle East which lack the water-proof integument of other Arthropoda (Cloudsley-Thompson, 1956a, 1959a, 1965a).

Scorpions and Solifugae often make deep excavations in the ground, the latter closing their burrows with a plug of dead leaves which no doubt helps to ameliorate conditions within. Scorpions of the family Scorpionidae are the best diggers and their enlarged claws or pedipalps are probably specially adapted for this purpose. *Scorpio maurus* whose range extends from the Atlantic to India, has been known to dig to a depth of 75 cm. The most important scorpion family is the Buthidae of which the North African *Buthus occitanus* is a well-known example. In less extreme deserts, buthid scorpions are usually found in shallow scrapes under rocks which they dig with their claws and legs, but *Leiurus quinquestriatus* and *Parabuthus hunteri* in the Sudan excavate deep burrows as do species of *Hadrurus* in America.

Many desert animals show adaptations for living in sand. Bristle-tails (Order Thysanura) form an important element of the desert fauna. They are extremely numerous in the Namib, for example, and take refuge at the base of tussocks of the grass *Aristida* sp. where they wriggle or almost swim with fish-like movements in the sand. At least one species of Hemiptera (bugs) has the same habit, and several species of tenebrionid beetles have become flattened and platelike in appearance, with short legs and the thorax and abdomen expanded into thin, wide plates with sharp edges. They burrow rapidly into the sand with alternate sideways movements. Some of these flattened beetles, which feed on the leeward side of the dunes, orientate themselves horizontally so that the smallest digging movement of the legs starts a cascade of sand from above which covers them very rapidly (Lawrence, 1959).

In the Sahara, too, many of the species associated with dunes and *ergs* show morphological adaptations for burrowing. Some are modified to swim through a loose substratum without making a hole, some excavate pits and others mine tunnels in more cohesive sand. A number of insects and Arachnida have enormous brushes of flattened hairs or bristles on the under sides of their legs which act like snowshoes and facilitate their movements through the sand (Pierre, 1958).

The lizards of sand dunes and *ergs* also show adaptation to the environment according to whether they are 'sand-runners' or 'sand-swimmers'. The former have the toes of the fore and hind limbs fringed with elongated scales. These presumably widen the surface which presses on loose sand, in the manner of snow shoes (Buxton, 1923). A modification that serves the same function is found in the duck-like webbed feet of the nocturnal gecko *(Palmatogecko rangei)* which lives among the sands of the Kalahari Desert. In this animal there is a complete webbing between the fingers and toes for support on loose sand. These various dilations of the toes are probably also of use in burrowing. Lizards that live in sandy deserts are usually extremely rapid in their movements. When not running, they stand alert with their heads held high and the front part of the body raised on the fore limbs so that they clear the hot sand. In motion, the tail is held well above the ground as a counterpoise. Such adaptations are found in a number of unrelated families from different parts of the world (Lawrence, 1959).

'Sand-swimmers' include the skinks as well as other lizards and snakes which exhibit several profound modifications for rapid burrowing in loose sand. The nose or *rostrum* is pointed or shovel-like and some species can dive head first into loose sand as though it were water. The nostrils tend to be directed upwards instead of forwards and thus are protected from the sand. As already mentioned (p. 84), they may be shielded by complicated valves, or are reduced to small pin-holes (Buxton, 1923). The eyes of the worm-like *Typhlops* spp. are overhung by large head shields. Lizards of the genus *Mabuya* may have the lower lid much enlarged with a transparent window in it so that the eye can be closed without impeding sight, an arrangement carried to an extreme in *Ablepharus* spp. where the lower lid is fused with the rim of the reduced upper lid (Gadow, 1901). The ear opening is also either small and protected by fringes of scales, or may even be abolished in certain reptiles. Desert lizards and snakes often have widened bodies for burrowing by lateral and vertical movements instead of ploughing forward into sand. Helical side-winding is found in snakes such as the American side-winder *(Crotalus cerastes)* and horned vipers of the genus *Aspis* in the Great Palaearctic Desert. These snakes progress by lateral loops of the body which cause them to move obliquely. (See discussion in Cloudsley-Thompson, 1971a).

Mammals may also show adaptations for life in sandy places. The camel, having lost all except two of its toes in the course of evolution, is unable to recover them but has increased the surface area of its feet by developing fleshy pads which do not sink into the sand. Its eyes are protected by its long and abundant lashes and it can close its nostrils at will to prevent the entrance of sand. In the saiga antelope *(Saiga tatarica)* the nostrils are turned back (p. 93) so that sand is excluded during grazing (Fig. 19). The bipedal gait of jerboas and kangaroo-rats, like that of marsupial kangaroos and certain desert lizards is an adaptation for speedy locomotion in open country (Cloudsley-Thompson and Chadwick, 1964).

Animal coloration

Most of the inhabitants of deserts are either black or pale in colour resembling their background. In many desert mammals the undersurface is very pale or quite white. Desert species and subspecies differ from their near relatives from other environments just as much in their pale ventral surface as in their buff or sandy backs. Moreover, the pale ventral area is often extended over the flanks, in desert forms, to a greater extent than it is in related species from other habitats. The same phenomenon occurs in spiders, centipedes, woodlice, insects, lizards, snakes and birds. Furthermore, desert-dwelling animals are not coloured fawn, brown, cream or grey indiscriminately. There is often a very close similarity between the creature and the soil of the particular type of desert on which it is living. For example, three dark subspecies of rodents (*Citellus grammurus tularosae, Perognathus intermedius ater* and *Neotoma albigula melas*) from dark, isolated larva beds in New Mexico contrast sharply with closely related species and subspecies that inhabit light soils and white gypsum sands in the vicinity (Sumner, 1921). Again, one of the crested larks (*Galerida theklae*) is represented by a dark race in the northern parts of Algeria which are not desert; in the semi-deserts which fringe the Atlas mountains southwards this is replaced by a paler race and, in the northern Sahara itself, a still paler race is found which is so wonderfully adapted to the soil that it is easily overlooked if the birds do not happen to be on the wing or singing. Between Laghouhrat and Ghardaia in Algeria, a reddish form is found and its distribution corresponds fairly closely with the reddish, stony desert. Similar examples could be quoted from many other species of birds and mammals (Cott, 1940). Although the theory that desert coloration in protection against predation has not been accepted by every student of the subject (Buxton, 1923), observations on the birds of the Namib Desert reveal no reason to reject it (Willoughby, 1969).

In addition to desert coloration, blacks and whites are also common. The occurrence of blackness in widely separated groups of animals is remarkable. Examples include ravens, wheatears, tenebrionid and scarab beetles, bees, flies and so on. Black is the most conspicuous colour against a background of sand, and can be considered as a warning coloration in distasteful or poisonous animals. Its prevalence among such desert animals may result from Müllerian mimicry, in which a number of different species, all possessing aposematic or warning attributes, resemble one another and so become more easily recognised. Numerical losses are reduced in teaching predators to avoid a common warning colour and the adoption of a common advertisement simplifies recognition. In Batesian mimicry, on the other hand, a relatively scarce, palatable and unprotected species, such as a bee-fly (Bombyliidae) resembles an abundant, relatively unpalatable or well-protected species such as a bee and, on account of its disguise, is ignored by potential enemies. In nature, Batesian and Müllerian mimicry tend to merge into one another as model and mimic become relatively more or less distasteful (Brower and Brower, 1972).

Hamilton (1973) argues that the black colours of desert animals must have a thermal significance. It has been found that transparent insect wings reflect more infra-red solar energy than black elytra, so the function of the black pigment cannot be to provide protection from radiation. Moreover, infra-red radiation can hardly be responsible for visible colour differences. Colour is probably relatively unimportant thermally because a high proportion of energy is transmitted at infra-red wavelengths. Furthermore, if black cuticle, feathers, or hair should become very much hotter than the environment, the excess heat would be removed by conduction, convection and, to some extent, by radiation, unless the heat were conducted inwards. This is not the case, because various insulating devices mitigate against it. (See p. 203).

The struggle for existence is especially severe in desert regions. Poisonous species tend to be more dangerous than their relatives from less arid environments, speedy animals faster, senses more acute. In a similar way, natural selection has played unusually heavily on the colours of desert animals. Cryptic species are exceptionally inconspicuous and Müllerian mimicry extremely common. Colours without marked adaptive significance are seldom, if ever found (Cloudsley-Thompson, 1964a; Cott, 1940).

6 STEPPE

Temperate glasslands occupy the interiors of continents in temperate regions with hot summers, cold winters and low annual rainfall. Only on the shores of lakes and along the banks of rivers is there usually sufficient moisture for trees to grow. Everywhere else, the land is covered with a rolling carpet of short grass. Such conditions prevail in the steppes of Eurasia, the North American prairies, the South American pampas, the South African veld, and the Australian downlands. In this book, however, sub-tropical steppe has been treated with savannah.

The world's largest area of temperate grasslands is the steppe of Europe and Asia, a vast plain stretching from Hungary through southern Russia to China in the east. Northward, it merges with deciduous or coniferous forest: to the south it is bordered by the Black Sea, the Caspian and the deserts of Central Asia.

The steppe lands of Europe and Asia are of particular interest because they were the original home of the nomadic Caucasian race, which later radiated throughout the world. Man has lived nomadically in the Eurasian steppe for longer than in any other part of the world. The fierce extremes of the climate have helped to engender bold and hardy warriors, the monotony of the landscape to encourage thinkers and dreamers who turned from the blank plain and empty sky to inward vision.

These steppes are also the areas from which the domestic horse has been derived. Apparently there have been at least three separate domestications: (a) a light weight, Indo-German variety has been derived from the tarpan *(Equus caballus gmelini)* which, until it was exterminated about 1880, lived in herds in the southern Russian steppe, or from some nearly related wild horse. The Arab is the purest breed of tarpon-like horses now in existence. (b) a western European stock of unknown ancestry from which have arisen modern draught horses. (c) a Mongolian race derived from the Mongolian wild horse *(E. caballus przewalskii)*. The mustangs of Mexico and California are feral descendents of horses imported by the Spaniards after the conquests of the sixteenth century. Small in size, hardy and sure-footed, they were later domesticated again and crossed with other breeds.

Steppe is also the home of several kinds of wild ass. The onager *(Equus onager)* of Central Asia is the wild ass of the Bible. It is white, with a large yellowish area on its sides, and a black mane and tail. The kiang *(E. kiang)*, which inhabits the high desert plateaux of Tibet, Ladak and Sikkim, is deep reddish-brown, with white legs, and is more solitary in its habits. The chigetai or kulan *(E. hemionus)* of southern Russia and Mongolia, is smaller, brownish or sand coloured.

The greatest cereal-producing regions of the world are situated in the steppes. It is no accident that the two most powerful nations between them contain most of the temperate steppe lands of the earth which, in turn, have gained their rich black earth at the expense of the deserts that lie beyond their southern borders.

Climate

The warm temperate zone contains various rainfall regimes as follows: (a) Rain at all season, evenly distributed; (b) Periodic rains with a maximum in spring and early summer; (c) Periodic rains with a maximum in winter; (d) Constant drought. Of these, (a) and (c) correspond with eastern margin sub-tropical and western margin sub-tropical (Mediterranean) climates respectively, and support temperate forest; (b) is found in steppe land and (d) in desert. Steppe grasslands also occur in cool temperate and cold continental climates (Appendix 1) characterised by summer rain that is insufficient to support coniferous forest.

The boundary between grassland and desert is difficult to demarcate, both in temperate, and in cold, or sub-arctic zones. Climate is clearly transitional, and local conditions determine plant associations. Tongues of forest may penetrate far into steppe along the banks of rivers, while strong winds may prevent the growth of trees, even in maritime regions which are well-watered.

Several qualities of cold continental climates favour the development of steppe. The cold of the winter may be phenomenal – -51°C (-60°F) is regularly experienced at Verkhoyansk in Siberia and temperatures as low as -68°C (-90°F) have been recorded. Indeed, the lowest temperatures in the world occur in this zone. In contrast, the summers are really hot. The means at Tobolsk for June and July are 15°C (59°F) and 18.5°C (65.5°F) respectively, while maxima of 32°C (90°F) are recorded almost every year. Added to these genial temperatures is the benefit of long hours of sunshine, whose effect on the vegetation is surprising (Miller, 1965).

So effective is the barrier of the western Eurasian mountains that summer rainfall is usually less than 50 cm (20 in.). Consequently, in central Asia, steppe degenerates into cold desert. The winters are so cold that the air is incapable of holding much moisture and anti-cyclone conditions in any case make precipitation unlikely.

Vegetation

Temperate steppe grassland represents a habitat in which, for climatic reasons, conditions are generally somewhat adverse to plant growth. In this, steppes contrast with savannah where, at the time of the rains, there is a brief season of luxuriant growth. Grassland is often the dominant association even where climatic conditions would otherwise favour forest were this not inhibited by strong winds, fire, or by human activities.

Cool temperate grasslands are found where the annual precipitation is less than 25-50 cm (10-20 in.). Rain falls early in summer and there is a considerable annual range of temperature. The heat of late summer scorches up the grass and causes annual prairie fires whose ashes add fertility to the rich black soils. This grass is well suited to stock raising and, in parts where rainfall is higher, is replaced by wheat and other cereals, which thrive under conditions of a moist spring followed by a hot, dry summer.

The vegetation of the steppes of central Asia is discussed by Kachkarov and Korovine (1942). The flora shows many of the characters of desert vegetation with regard to rooting systems, and other xerophyllic characters, and is much influenced by soil, agriculture and grazing. There is no doubt that much steppe, like savannah, has been transformed into desert by human mismanagement (Kassas, 1970).

Steppes are even more varied in type than are savannahs: they are found in sub-tropical as well as in temperate climates, and both at low and high altitudes. The vegetation does not attain the luxuriance of savannah grassland, but all gradations are found from dense short grass to semi-desert. With the grass are found many kinds of flowering plants, species with bulbs and corms, poppies, thistles, and low woody plants, such as *Artemisia* spp., which may predominate. In many steppe areas, there are stretches where the soil is rich in salts and supports only a sparse growth of halophytes (Hesse, Allee and Schmidt, 1951). The dominant grass of Eurasian steppes is the meadow grass *(Poa pratensis)*. This has been imported into America, where it is known as Kentucky blue grass, and is popular in horse-raising areas as a staple feed.

Before Europeans came to North America, bringing their grazing stock with them, two-fifths of the continental United States and much of Canada were grassland. No adequate description has been written of the original western range, and probably the most comprehensive account is that of Shantz and Zon (1924). The range consisted of prairies, plains, deserts and mountains, of which the prairie was the most productive. Dominant grasses of the tall-grass prairie included the bluestems *(Andropogon* spp.*)*, needlegrasses *(Stipa* spp.*)*, switchgrass *(Panicum vergatum)*, and many species palatable to sheep and cattle. Herbs, trees and shrubs of lower stature also occur. Tall-grass prairie was aptly named because the flower-stalks of *A. gerardi* and *Sorghastrum mutans* reach heights of 1.5-2.5 m.

Short-grass prairies covered an area between the tall-grass prairies on the east and the Rocky Mountains on the west. The principal species of grass were the gramas *(Bouteloua* spp.*)*, buffalo grass *(Buchloe dactyloides)*, bluestem wheatgrass *(Agropyron smithii)* and needle-and-thread *(Stipa comata)*. These also provided nutritious foliage. Pacific bunchgrass occurred in Montana, the Pacific northwest, and in central California. The characteristic species were Idaho fescue *(Festuca idahoensis)* and bearded bluebunch wheat-grass *(Agropyron spicatum)* in the north-west and California needlegrass *(Stipa pulchra)* in the great central valley of California (Costello, 1964). The semi-desert grasslands of Arizona and New Mexico and the colder Great Basin desert of Utah, Nevada and Idaho were discussed in Chapter 5. Open forests of pinon pines *(Pinus* spp.*)* and junipers *(Juniperus* spp.*)* characterised the foothills of many mountain ranges from Colorado to Oregon and California. As a result of overgrazing and range mismanagement, the forage was depleted and shrubs invaded many parts of the range. Erosion was rampant between the middle of the last century and the beginning of this one: the whole biome is now largely artificial.

Prairie has rather clear layers or strata, reflecting the different statures of common grasses and herbs. The different types of prairie are best identified by the use of plant indicators on lightly or non-grazed areas, for grazing animals can greatly influence the composition of the flora.

Fauna

Mammals

The fauna of steppe and other grassland appears to have few, if any, elements peculiar to itself. Almost all its species also inhabit the forests and deserts to the north or south, and seldom show adaptations that are peculiar to open country. In response to the lack of cover and shelter, and the extremes of climate experienced, mammals tend either to be gregarious and speedy, or else to have evolved burrowing habits. Consequently, the dominant forms are either social ungulates or rodents.

Characteristic of the Asian steppes are the saiga antelope *(Saiga tatarica)* (Fig. 19), the maral stag *(Cervus elaphus)*, the goitered antelope *(Gazella subgutturosa)*, the wild ass *(Equus onager)* of which the herds are led by an old stallion and the kulan *(E. hemionus)*. Other large mammals such as hares, boars, roebuck *(Capreolus* spp.*)*, polecats *(Mustela putorius)*, weasels *(M. nivalis)*, badgers *(Meles meles)*, foxes *(Vulpes vulpes)*, gluttons *(Gulo gulo)*, wolves *(Canis lupus)* and the lynx *(Lynx lynx)* are less typical and are often found also in forest and desert.

For centuries the steppe grasslands of North America were the home of bison *(Bison bison)* and pronghorns *(Antilopcapra americana).* A century ago, the bison was the chief grazing animal in the centre of the continent but, in a comparatively few years it was entirely replaced by cattle and sheep, as well as by other kinds of farming (Elton, 1958). The structure and competition of the prairie vegetation also changed, but the 'buffalo bird' *(Molothrus ater)* of the pioneer days remained as the 'cow bird' of modern farm pastures (Roe, 1951). It perches on the backs of bison and cattle, picking ticks from their long hair or swooping down to snatch at grasshoppers flushed by the grazing of its host.

Pronghorns are the fastest mammals of North America: speed is their main defence against predatory enemies, such as wolves *(Canis lupus)* and coyotes *(C. latrans).* In small herds, they range over the dry prairies in summer, eating grass, sagebrush *(Artemisia tridentata)* and other plants. Later in the year they form larger herds, of 100 individuals or more, and move south for the winter. The calves are born in April or May when the grass is growing most strongly (Matthews and Carrington, 1970).

In a study of the distribution of small mammals in continuous forest, aspen groves, riparian woodland and prairie-forest transition of Minnesota and North Dakota, 13 taxa were collected. These could be divided into three groups on the basis of habitat selection. Of the forest species of white footed mice, *(Peromyscus maniculatus gracilis)* was the most restricted and occurred only in or near coniferous forest. *P. leucopus noveboracensis* was found in coniferous woodland, deciduous and river-bottom forest and nearby groves. The vole *(Clethrionomys gapperi)* had a similar distribution, but was also found in some small aspen groves with grassy ground cover. Chipmunks *(Tamias striatus)* were collected in the continuous forest and nearby large groves; the species *Eutamias minimus* and *Synaptomys coopri* only in continuous forest. The prairie species of white-footed mouse *(P. m. bairdii),* grasshopper mouse *(Onychomys leucogaster),* and meadow mouse *(Microtus pennsylvanicus)* were found essentially in grassland situations, but *P. m. bairdii* occurred also in aspen groves and riparian forests from where *P. l. noveboracensis* was absent. *M. pennsylvanicus* was a permanent resident of young groves with grassy undercover, while shrews *(Sorex cinereus* and *Blarina brevicauda)* and jumping mice *(Zapus hudsonius)* were found in all habitats. *Sorex arcticus* was also present in moister habitats throughout the transition (Iverson, Seabloom and Hnatiuk, 1967). The fauna of the prairie and forest, containing both woodland and grassland species, reflects the state of flux in the plant associations. There is an inverse correlation between the abundance of small mammals in grassland habitats and the amount of herbaceous vegetation (Mossman, 1955).

Birds

P.S. Nazarov (1886, cited by Haviland, 1926) studied the distribution of birds in the region of Khirghiz and recognised four steppe formations, each with a characteristic avifauna, especially of eagles and larks. These were as follows:

(a) The parkland region, including the forested slopes of the Urals. Here the grassland is rich, comparatively well watered and interspersed with deciduous woodland. The fauna is transitional between that of forest and true steppe and includes the common squirrel *(Sciurus europeus)*, flying squirrel *(Pteromys volans)*, marten *(Martes martes)*, brown bear *(Ursus arctos)* and sousliks or ground squirrels *(Citellus* spp.*)*. Characteristic birds are the imperial eagle *(Aquila heliaca)*, sparrowhawk *(Accipiter nisus)*, hobby *(Falco subbuteo)*, skylark *(Alauda arvensis)*, great bustard *(Otis tarda)*, woodcock *(Scolopax rusticola)*, and many woodland species.

(b) Stipa steppe, in which *Stipa pennata* and *S. capillata* are the dominant species of grasses. Trees appear along the valleys of the rivers and there is abundant vegetation in spring. Drought sets in at the end of June and animal life is then concentrated around the lakes. Characteristic mammals here are the saiga antelope *(Saiga tatarica)*, marmots *(Marmota* spp.*)*, the jumping-rabbit *(Altaga jaculus)* and corsae fox *(Canis corsae)*: birds include the rosy starling *(Pastor roseus)*, crane *(Megalornis grus)*, McQueen's bustard *(Otis macqueeni)*, sandgrouse *(Pterocles* spp.*)*, steppe eagle *(Aquila nipalensis)* and larks *(Alauda tatarica* and *Melanocorypha sibirica)*.

(c) Artemisia steppe. This is even more arid: grass grows poorly and the land is clothed with xerophytes such as *Artemisia* spp. and a few thorny bushes. The eagle *Aquila glitchi* is characteristic, as are larks *(A. brachydactyla* and *M. calandra)*, McQueen's bustard and sandgrouse.

(d) Sandy steppe which merges into desert where jerboas *(Dipus sagitta)*, gerbils *(Gerbillus* spp.*)*, hedgehogs *(Erinaceus* spp.*)*, sand hamsters *(Cricetus arenarius)*, eagles *(A. bifasciata)* and shore larks *(Otocorys alpestris)* are characteristic.

Although they refer only to a small and peculiar part of the steppe region, Nazarov's observations are representative of the changes in vegetation and fauna which occur in this whole geographic region.

In the absence of trees and rocky crags, many birds that would normally breed in such places instead make use of mud ravines or nest on the open ground. Some species may undergo population explosions and then emigrate in vast numbers. An example is afforded by Pallas's sandgrouse *(Syrrhaptes paradoxus)* which occasionally irrupts into Europe. In 1863 it arrived in great numbers, penetrating as far as Great Britain, Ireland, the Shetlands and Faroes. Although it bred in Britain and Denmark, it did not establish itself. The last big invasions took place in 1888 and 1908. Rough-legged buzzards *(Buteo rufinus)* are also to be included among emigrants from the steppes of central Asia (Heape, 1931).

The dry steppes and semi-desert regions of Europe and Asia possess a rich semi-aquatic or lacustrine fauna centred round lakes and rivers. Steppe lakes, fed by the melting snows, are often seasonal and dry up to marshes in summer. Nevertheless, in summer they may support a rich and varied fauna of birds such as herons, storks, bitterns, ducks, grebes, avocets, plovers, stilts, harriers and warblers (Haviland, 1926).

Reptiles and amphibians

Open country is especially favourable to poikilothermic or ectothermal animals which thermoregulate largely by behavioural means. By basking in the sun, reptiles are able to raise their body temperatures to levels corresponding to those of homeotherms. Excess temperatures are avoided by altering the orientation of the body so that the long axis becomes parallel to the rays of the sun, by climbing trees, or retreating into cool underground shelters (Cloudsley-Thompson, 1971a; 1972b). Behavioural regulation of temperature is supplemented by physiological mechanisms such as panting, salivating, producing metabolic heat and shunting blood from one part of the body to another.

Lizards and snakes are present on grassland in large numbers Tortoises and terrapins, such as *Terrapene ornata* in western North America, may be plentiful locally. Sandy soils readily permit burrowing, and reptiles are numerous in such terrain. Grasshoppers, termites and other insects form the staple food of lizards: they are also devoured by snakes along with mice, rats and other small mammals.

Reptiles hibernate in cold weather. Although not usually found in arctic regions, snakes occur between latitudes 67°N in Europe (where the winter cold is ameliorated by the effects of the Gulf Stream), 60°N in Asia, 52°N in North America, and 44°S in South America and Queensland (Curran and Kauffeld, 1937). Daily ambient temperatures must, however, be sufficiently high in summer to permit activity; the annual period during which such temperatures occur must be long enough to allow breeding to take place, and to enable the young and adults to acquire nutritional reserves for their long winter hibernation; and there must be adequate niches available for hibernation (Hock, 1964a). Only ovoviviparous species can survive in cool regions. This is because eggs would be unable to develop on account of low air temperatures and the short annual season of activity.

Amphibians are scarce in the steppe and all are burrowers. During the spring and early summer they lay down stores of fat to tide them over the drought of late summer and the winter cold. The molluscan fauna is likewise impoverished and the only species are those that can endure long periods of quiescence. In the North American prairies, they are limited mainly to the tree-grown banks of rivers. The pampas are almost lacking snails as are the Australian downlands.

Insects

The steppe is a paradise for Orthoptera. Grasshoppers of all sizes, colours and powers of flight, from great locusts to little crickets, are plentiful. Many of them exhibit 'flash' coloration, but are otherwise inconspicuous. By no means all steppe insects are cryptically coloured, however, and some of the Heteroptera, Hymenoptera and Diptera are highly conspicuous.

Migratory locusts of various species inhabit the steppe lands of all continents, being more common in semi-arid regions. Examples include *Locusta migratoria,* widely distributed in the warmer parts of the Old World, and the Rocky Mountain locust *(Melanoplus mexicanus).* Locusts and grasshoppers are particularly important in the ecology of steppe land because they transform the hard summer grass into easily digested materials and so indirectly provide nourishment for a large number of animals including scorpions and Solifugae, lizards, snakes and turtles, hawks, buzzards, owls, storks, ravens and a host of others. In years when locusts are especially numerous, the rosy starling *(Pastor roseus)* is attracted far from its normal range in pursuit of the swarms, and grain-eating birds feed them to their young (Hesse, Allee and Schmidt, 1951).

Ants and, in warmer regions, termites, likewise form the bases of important food pyramids and are eaten by many mammals and birds which dig up their nests to obtain the larvae and pupae. Some species of ants regularly store quantities of grain to support their colonies in times of drought. Harvesting ants of the genus *Messor* do so in the Mediterranean region and, in North America, species of *Pogonomyrmex.* Dung-beetles (Scarabaeidae) are common in steppe grassland whose ungulate fauna provides them with abundant food.

The total number of insects present may be impressive. According to Bird (1930) they may number 9,500,000 per acre in spring when many insect-eating vertebrates are still hibernating and before flocks of migrant birds arrive. Numbers are reduced to only 1,000,000 per acre, however, in late June.

Characteristics of steppe animals

The animal life of the steppe grassland depends, of course, upon the nature of the vegetation. Much of this consists of hardened stems, dried, stiff and often thorny stalks, seeds and roots. To each such food, strong masticating apparatus is necessary. Grasshoppers and ants possess powerful mandibles, rodents and ungulates have front teeth adapted for clipping the vegetation and strong molars with broad, roughened crowns for grinding it. These rear teeth are endowed with the capacity for continued growth, as they are steadily worn away. Grain-eating birds, such as sandgrouse, grind their food by means of their muscular gizzards. Bird (1930) discussed the nutritional interrelationships of the fauna in the aspen parkland region of Manitoba, showing how animals that feed on grass and grain constitute the bases of the various food chains in the region.

Steppe grassland is characterised by hot, dry summers when many animals aestivate, and cold winters when snails, insects, spiders, amphibians, reptiles and small mammals hibernate. Most birds migrate to warmer climes. Burrowing rodents, such as hamsters (*Cricetus* spp.) and sousliks (*Citellus* spp.), remain active, feeding on supplies of food stored in their burrows. *Ochotona dauricus* of the Asian steppes gathers stacks of grass weighing up to 10 kg in the vicinity of its burrows.

Larger mammals are less affected by cold and have thick winter coats of fur. They scrape away the snow to feed on dry grass, moss and lichen, or nibble twigs and dry leaves. Many of them migrate, driven, perhaps, not so much by shortage of food as of water when everything is frozen. As already mentioned, they tend either to be mobile, social ungulates or else the carnivores that prey on them.

7 TEMPERATE FOREST

Temperate forests take many forms. In cooler latitudes the trees are mainly deciduous, but evergreens predominate in warmer regions. The three main types of temperate forest are as follows:-

(a) *Temperate deciduous forest.* Originally, this covered eastern North America, Europe, part of Japan, Australia and South America.

(b) *Moist temperate coniferous forest.* This is quite distinct from 'taiga' (Chapter 8). It is found in western North America from California to Alaska and in the Mississippi delta, where temperatures are relatively high with a small seasonal range. The humidity is generally high with precipitation from 45-380 cm (30-150 in.) often supplemented by fog. Conifers such as red woods *(Sequoia* spp.*)* and spruces *(Picea* spp.*)* dominate.

(c) *Broad-leaved evergreen forest.* This is found where moisture is high, and differences in temperature between summer and winter are less marked than in regions of temperate deciduous forest. It is characteristic of Florida and of central and southern Japan. Live oaks *(Quercus virginiana),* magnolias, hollies *(Ilex* spp.*)*, bays, and sabal palms *(Sabal palmetto)* are typical (Odum, 1971). Very little of the temperate forest of the world remains in its natural state. Most has been profoundly modified by human activity.

Climate

Warm temperate climates are situated in the latitude of the oscillating front of divergence which divides the spheres of influence of the trade winds and the westerlies. These climates are characteristically transitional: for part of the year they are 'tropical' in their consistency but, during the remainder, the weather is typically 'temperate' in its changeability. The summer is therefore 'continental' on western margins (p. 108) and 'marine' on eastern margins. Conversely, winter is 'marine' on western margins and 'continental' on eastern margins. Western margin type is found around the Mediterranean Sea, along the west coast of North and South America (California and Chile), on the south Australian littoral and in the neighbourhood of Cape Town. Mediterranean climates, therefore, tend to be hot and dry in summer and comparatively warm but wet in winter. Although winter is the rainy season, it is by no means cloudy or damp.

Eastern margin warm temperate climates are characterised by mild winters with mean temperatures about 10°C (50°F) while the summers are hot and humid. Rainfall is not excessive, but adequate, and well distributed over the year. In winter it is chiefly of the cyclonic type, occurring as light showers or prolonged drizzles. In summer, on the other hand, it tends to occur in heavy downpours.

Cool temperate marine climates show a smaller annual range of temperature and a higher humidity and rainfall than continental climates. Rainfall is evenly distributed throughout the year with a tendency to a maximum in winter and the seasons grade imperceptibly into one another.

Vegetation

The belt of frost-resistant forests that occupies the temperate zones of the northern hemisphere south of the boreal forest and taiga has no true equivalent south of the equator. It consists of many disconnected populations of trees, separated by long established barriers of ocean and desert. In both hemispheres, however, closely related species of the same genera occupy similar ecological niches. The main forest trees are pines, oaks, beeches and species of *Eucalyptus*.

Despite their wide distribution and diversity, pines *(Pinus* spp.*)* are generally characteristic of coarse dry soils such as sands, gravels and outcrops of rock. They owe their dominance to the fact that they are able to regenerate rapidly after burning. They have deep roots and consequently do not grow well on frozen or poorly-drained soils, but can withstand hot, dry conditions both in the tropics and in northern continental climates. As already mentioned, *P. ponderosa* is the principle hard pine of western North America where *P. contorta* and *P. jeffreyi* are also of major importance. On the eastern side of the continent, red pine *(P. resinosa)* and the four southern species *(P. taeda, P. palustris, P. elliotii* and *P. echinata)* are the major hard pines.

The dry mountain ranges of southern Europe and Asia Minor are populated by many species of pines although they have been severely limited by centuries of grazing and burning. The most important species are *P. nigra,* which is found from the Pyrenees to Turkey and Cyprus, the maritime *P. pinaster* and *P. halepensis, P. brutea* and *P. pinea.* In Asia are found *P. sylvestris, P. cembra* and *P. longifolia.* Other conifers of temperate forest include cedars *(Cedrus* spp.*)*, cypresses etc.

Members of the genus *Quercus* are the angiosperm equivalents of the pines – a widely distributed group of deep-rooted, zerophytic trees that occupy dry sites from the southern edge of the boreal or northern forest, well into the tropics. In the north temperate zone they are deciduous but, south of the equator they are frequently evergreen. In North America *Q. rubra, Q. velutina* and *Q. alba* are among the most important species, *Q. robur* and *Q. pitraea* in Europe and eastern Asia. The Mediterranean forests are dominated by the cork oak *(Q. suber)*, the holm oak *(Q. ilex)*, *Q. macrolepsis* etc. *Q. mongolica* is widely distributed throughout southeastern Siberia, Mongolia, Manchuria and Korea. Other hardwoods include beeches *(Fagus* spp.*)*, maples *(Acer* spp.*)* and many other genera including limes *(Tulia* spp.*)*, chestnuts *(Castanea* spp.*)*, sycamores *(Plantanus* spp.*)*, alders *(Alnus* spp.*)*, ashes *(Fraxinus* spp.*)* and elms *(Ulmus* spp.*)*. (For further details, see Spurr, 1964).

The dominant characteristic of Mediterranean (western margin) climates is the marked rhythmic recurrence of rain and drought, which is naturally reflected in the growth of plants. Since the rain falls in winter when temperatures are too low for vigorous growth, and the summer is a time of drought, the season of greatest vigour are autumn and spring, when temperatures are moderate and there is yet adequate rain.

Where conditions are most favourable and rainfall heaviest the natural vegetation consists of evergreen woodland with pine, cedar and evergreen oak. Deciduous oaks are found when drought is less severe. In western Australia, the forests are of *Eucalyptus* spp.

As in Mediterranean climates, the winters of eastern margin warm temperate regions are never cold enough to inhibit entirely the growth of plants, and many species of conifers such as cypresses *(Cupressus* spp.*)*, shrubs (e.g. laurel, *Kalmia* spp.) and evergreen oaks are common to both climates. The absence of a dry season, however, enables tree ferns, bamboos, lianas, magnolias and other handsome species to survive. These are excluded from western margin climates. The regular rainfall supports a forest vegetation usually of broad-leaved evergreens but sometimes of deciduous or coniferous trees. There are, however, considerable areas, such as the Argentine pampas, which support a grassland vegetation although precipitation is adequate for forest growth. They have been discussed in Chapter 6.

Broad-leaved deciduous forest, characteristic of western margin climates, extends for a considerable distance inland. It reaches its finest development in the semi-oceanic climate of western Europe where oak *Quercus* spp.), ash *(Fraxinus excelsior)*, beech *(Fagus sylvatica)* and maple *(Acer campestre)* are the chief forest trees, while elm *(Ulmus* spp.), chestnut *(Aesculus hippocastanum)*, sycamore *(Acer pseudoplatanus)* and lime *(Tilia platyphyllos)* are common. It is a feature of these, as of coniferous forests, that they often consist of almost pure stands of a single species – a fact which adds considerably to their economic value. In North America, the equivalent forest is largely coniferous and does not extend beyond the Cordillera.

According to Yapp (1953), there are four types of woodland, other than obvious plantations of exotics, above the 305 m (1,000 ft.) contour in the Lake District of the British Isles. These are pure sessile oakwood, birchwood, alderwood and a mixed association of oak-ash-birch.

The climax forests of central Europe, particularly in Switzerland and adjacent mountain areas of France, Germany, and Czechoslovakia, are dominated by Norway spruce *(Picea abies)*, silver-fir *(Abies alba)* and beech *(Fegus sylvatica)*. These long-lived, tolerant species nevertheless form an unstable climax with spruce replacing fir, fir replacing spruce and similar changes involving beech. This phenomenon is known as 'alternation of species'.

The boreal forest of western North America forms a community characterised by distinctive species that tend to be restricted in number. Development is patchy in the north, reflecting the nature of the glacial plain upon which the forest grows. A mature forest composed of sitka spruce *(Picea sitchensis)* and western hemlock *(Tsuga heterophylla)* is eventually followed by marsh formation, which may eventually be swamped to produce pit-ponds and moss-bogs ('muskegs'). Given time, sphagnum invades coniferous forests, bringing about the accumulation of upland peat (Spurr, 1964).

Meadows are common, as are ponds caused by the activities of beavers *(Castor canadensis)* or by cirque moraines.* If shallow, these become dense willow thickets. On moderate slopes, lower subalpine forest may be dense. The trees include conifers such as *Tsuga* spp., *Picea glauca* and *P. engelmanii, Abies lasiocarpa, Populus tremuloides* or *Pinus contorta* which tend to produce even-aged stands. Such forest grades into 'taiga' (Chapter 8).

Most of the temperate forests of the Palaearctic region have been destroyed by man. It is, therefore, perhaps of academic interest only to consider the composition of their climax vegetation. Nevertheless, this must be understood if one is to assess the effects of agriculture and other human activities on the status of the biome. Although such considerations will be treated in other volumes of this series, two long-established but relatively unnatural sub-climaxes merit special mention; viz. chalk downs and heathlands.

Man's role in changing the face of the earth is very great. The chaparral belt in the Sierra Nevada of California, the maquis of the Mediterranean, the manuka scrub of New Zealand, and vast areas of deforested land in China result from indiscriminate burning, logging, farming and domestic grazing.

Forests form the natural vegetation in many parts of the world where precipitation exceeds evaporation. Typical forest soils, therefore, usually comprise a full range of horizons (p. 25). The development of the profile may continue over many thousands of years. For instance,

---

* Cirque moraines – circular deposits of debris brought down mountain sides by glaciers.

similar parent materials in the northern portion of the lower peninsula of Michigan can be dated from their position in relation to old levels of the Great Lakes. One series, exposed for 3,500 years, shows only slight weathering and profile development. Even on sand, lime is available within 30 cm of the surface, to an extent favouring the growth of northern white cedar *(Thuja occidentalis).* On similar sand soils, exposed for 8,000-10,000 years, the lime has been leached to a depth of 1 m or so, and clearly segregated A and B horizons have been formed. On adjacent clay loams soil profiles are still relatively immature after 10,000 years (Spurr, 1964). When the forest and its soils have been destroyed, they cannot quickly be restored.

Chalk downs

Rendzina soils are typical of chalkland and are formed *in situ* by weathering of the underlying parent limestone. They support a sward – an expanse of herbaceous vegetation – rich in plant species, which passes in a succession of shrubs to a climax of mature woodland. It is doubtful if any chalk downland today supports natural vegetation unaffected by human activities. Large areas of the downs of Europe have for centuries been grazed by sheep which, by continual tramping and nibbling of the vegetation, killing seedlings of shrubs and trees, have arrested the development of the natural succession (Sankey, 1966).

The most common grasses of the sward, typical of different plant communities, include *Zerna erecta* and sheep's fescue *(Festuca ovina).* Other herbs which may be plentiful locally are the daisy *(Bellis perennis),* sedge *(Carex flacca)* and plantains *(Plantago media* and *P. lanceolata).* In less trampled regions buttercups *(Ranunculus bulbosus),* thyme *(Thymus drucei),* flax *(Linum catharticum),* rough hawkbit *(Leontodon hispidus)* and salad burnet *(Poterium sanguisorba)* are common. About 35 per cent of species show the rosette habit. By spreading outwards they prevent other plants from growing near them and thus eliminate competition.

The scrub is composed of shrubs or bushes intermediate in height between the herbs and the trees. These are often scattered, with well-developed sward between them, or they may be close together with little vegetation underneath. Hawthorn *(Crataegus monogyna)* and dogwood *(Thelycrania sanguinea)* are two of the most important species, although the former is often replaced by spindle *(Euonymus europaeus)* and wayfaring tree *(Viburnum lantana)* where the soils are shallower.

By far the most widespread climax of succession is beech *(Fagus sylvatica)* woodland, though a sub-climax of ash *(Fraxinus excelsior)* is often attained first, especially on north-facing slopes and on deeper valley soils. In some places yew *(Taxus baccata)* may replace beech as the climax and it is often found as an under-storey of the latter

(Sankey, 1966).

Salisbury (1952) has outlined the manner in which the original woodland has been changed by human activity to form the grassy swards of the chalk escarpments. These have been the grazing grounds of sheep since Neolithic man first brought agriculture to the British Isles.

The rich fauna of a chalk down reflects the diversity of plant species upon which it depends. Snails, such as *Helix pomatia* and *Pomatias elegans* are characteristic, as are blue butterflies *(Lysandra* spp.*)* and the beetles *Brachinus crepitans* and *Oedemera lurida.*

Moors and heathland

Over wide areas of western and north-western Europe, cool temperate forest is replaced by heather moors in which ling heather *(Calluna vulgaris)* is the dominant species, often mixed with species of *Erica,* related heaths and the crowberry *(Empetrum nigrum).* Although of a different family than the heaths, crowberries bear a striking resemblance to cranberries, bilberries, cowberries and so on.

As Newbigin (1936) points out, these plants are common associates of the lighter coniferous woods, especially of those in which birches are plentiful. The shrubs are comparable to those which form the undergrowth of sclerophyllous Mediterranean woodlands that undergo summer drought, and are often of the same genera. With the exception of *Calluna vulgaris,* which stretches across Greenland to Newfoundland and then south to Massachusetts, true heaths are confined to Europe. Consequently, heather moors are virtually limited to this continent, where they intermingle with and grade into wet moors with cotton-grass *(Eriophorum* spp.*)* and *Sphagnum* moss, where the characters of the vegetation are controlled by the presence of absence of peat.

The ecology of heathland has been described by Friedlander (1960) who points out that, although characteristic of sandstone beds and alluvial gravels, where podsols are produced by leaching, heathland is not a climatic climax. It is an unstable community whose species are kept in check by several factors, including the felling of trees, burning, grazing and trampling. Gimingham (1972) discusses the history and composition of heath communities and their micro-environments.

The most widespread dominant on dry heath is ling heather. Bilberry *(Vaccinum myrtillus)* often forms a subsidiary layer under the ling, but may become dominant, especially on higher ground. It is able to thrive on the same ground as ling because it uses a different part of the soil profile, thus reducing competition. It is a variable plant: it produces xeromorphic 'sun' leaves when it grows in exposed situations, such as burned patches, and 'shade' leaves when sheltered, as in woodland. On some moors, bracken *(Pteridium aquilinum)* is dominant; in others, bell heather *(Erica cinerea* or *E. tetralix).* Other common plants are sheep's sorrel *(Rumex acetosella),* plantains *(Plantago coronopus* and *P. lanceolata),* gorse *(Ulex europaeus)* and broom *(Sarothamnus scoparius).*

Fauna

Mammals

Conditions of life are very different in forest than they are in open country, and the fauna is correspondingly different too. Forest can exist only where rainfall is reasonably high. Consequently, the air is damp, the ground soft, and there is an abundance of juicy leaves and fruit to support the fauna.

The inhabitants of temperate forests do not show the prolixity and diversity seen in the fauna of rain-forests, but they enjoy a greater range of environmental adaptations than are found in open country. Non-arboreal mammals include long-tailed and short-tailed shrews *(Sorex* and *Blarina* spp.*),* the pine mouse *(Pitymys pinetorum),* red-backed mouse *(Evotomys gapperi),* and species of *Peromyscus* in America and of *Apodemus* in Europe. Such forms inhabit the litter layers of the floor and burrow into the soil beneath. Many insectivorous mammals of deciduous forest hibernate in winter. These include hedgehogs *(Erinaceus europaeus)* and bats. Shrews do not hibernate, however, because they have such a high metabolic rate that they must eat continuously in order to survive. Their metabolic rate drops when they sleep.

Other characteristic terrestrial mammals of temperate forests are red deer *(Cervus elaphus),* roe deer *(Capreolus* spp.*),* wild boar *(Sus scrofa),* badgers *(Meles meles),* foxes *(Vulpes vulpes* and *V. fulva),* wild cats *(Felis silvestris),* martens *(Martes* spp.*)* and lynx *(Lynx* spp.*).* Many of these inhabit subterranean dens. Seton (1909) lists thirteen species of mammals in Manitoba that are mainly arboreal, including squirrels, flying-squirrels and bats; eighteen terrestrial species and three that are mainly burrowers. The arboreal opposums *(Didelphys virginiana)* and porcupines *(Erethizon* spp.*)* of the Nearctic region are not found in the Palaearctic.

European temperate forests have been influenced so much by man in the interests of agriculture, game conservation and the preservation of rare species that it is difficult to reconstruct their original constitution. In olden time, bears and wolves were numerous in England, while wild cattle hid their young in coverts and grazed in the open. Even today, the wild boar ranges the forests of Europe and Asia. The tiger, found in the birch forests of southern Siberia, is absent from Ceylon and Borneo. Evidently it has only recently crossed the Himalayas into the tropical regions we tend to regard as its natural home.

Birds

The temperate forest biome has a rich avifauna. Woodpeckers, nuthatches, warblers, thrushes, hawks and owls find their food even in deep forest. As in the tropics, many other birds are to be found in the forest margins, including ravens *(Corvus corax),* the black stork *(Ciconia nigra)* and herons *(Ardea cinerca)* in Eurasia, and crows *(Corvus brachyrhynchos),* cardinals *(Richmondena cardinalis),* flickers *(Colaptes* spp.*)*, and brown thrushes *(Toxostoma* spp.*)* in America. Certain species are confined to temperate deciuous forests, others to coniferous ones.

Many parts of a tree are edible, but the highest concentrations of protein are found in the buds and seeds which, between them, provide a source of food throughout the year. Buds are often available throughout the year but are especially valuable in winter and early spring when other foods are scarce. Seeds and berries are the harvest of the autumn and winter, insects of the summer months. Many species of birds and mammals follow an annual pattern of feeding, eating mainly buds in spring, insects in summer when the young are being reared, and seeds in the autumn and winter (Matthews and Carrington, 1970). Such a cycle is typical of the hawfinch *(Coccothraustes coccothraustes)* and other Fringillidae of the Palaearctic region.

The hawfinch can exert a force of 30-40 kg at the tip of its beak to crack a cherry stone. It spends most of its life in the trees and is seldom seen on the ground except in winter when it feeds on fallen seeds. Many of the grosbeaks *(Pheucticus* spp.*)* of North America have similar feeding habits. Acorns provide the autumn food of pigeons *(Columba palumbus),* jays *(Garrulus glandarius),* woodpeckers *(Melanerpes formicivorus)* and the Eurasian nutcracker *(Nucifraga caryocatactes).*

Most insectivorous birds either feed on seeds during the winter, or else they migrate to areas where insects remain active. Tits (Paridae) and tree-creepers *(Certhia familiaris),* however, remain in deciduous forest throughout the year and survive the winter by finding the larvae and pupae of insects. They have such specialised habits and food preferences that several species can live together without competition. The most familiar tits in Europe are the bluetit *(Parus caeruleus),* the great tit *(P. major),* and the willow tit *(P. montanus).* In North America, the best known species is the black-capped chickadee *(P. atricapillus).*

Crows, rooks *(Corvus frugilegus)* and magpies *(Pica pica)* are omnivorous, and will readily feed on fruit, grain, eggs, insects, small birds, mammals and carrion. Their versatility and intelligence enables them easily to survive the winter without migrating.

Reptiles

The reptiles and amphibians of temperate regions are relatively scarce in numbers, both of individuals and of species, in comparison with their profusion and variety in tropical and sub-tropical areas. The viviparous lizard *(Lacerta vivipara)* and the slow worm *(Anguis fragilis)* are widely distributed throughout the Palaearctic deciduous forest region, as is the adder *(Vipera berus).* All three species are ovoviviparous, and give birth to living young. By basking in the sun during the day and hiding under cover at night, a reptile is able to maintain a body temperature continuously above that of the environment. This is probably necessary for the development of the eggs, as already pointed out (p. 113). The American garter-snake *(Thamnophis sirtalis),* whose range extends as far north as Alaska, is likewise viviparous, as is the European grass snake *(Natrix natrix).* This species, and the slow-worm occur as far as 63°30'N in Norway (Hock, 1964a).

Vipera berus can withstand -2.5°C for some time, while hibernating in southern Finland, and -4°C for a short while. Consequently, burrowing to a depth of 5-35 cm is adequate in normal years for vipers to survive the winter, and a depth of 50 cm in cold years. In northern Finland, lethal temperatures pentrate deeper than 100 cm, and this determines the northern limit of the viper (Viitanen, 1967).

Invertebrates

The fauna of the forest is more varied than that of steppe and prairie because the environment offers a wider range of conditions. The soil fauna is richer (Chapter 11) while that of the vegetation is divided between the foliage and the tree trunks. The insects on the foliage may be surface feeders, leaf-rollers, leaf-miners, or gall formers. Those of the tree trunks may inhabit cambium, cambium-wood, heart-wood, or they may dwell beneath the bark (Chapman, 1931). Forest entomology is a vast subject which lies beyond the scope of the present volume.

A single oak tree can harbour 50,000 caterpillars; a single leaf, more than 500 aphids. These insects, not plant-eating mammals, are the defoliators of forests. Leaf-chewers, such as caterpillars, are equipped with jaws adapted for biting, while sap-suckers, such as aphids, withdraw fluids through mouth-parts that operate like hypodermic needles.

It is not surprising with so many insects about, that temperate forests should be the paradise of spiders.

Seasonal breeding cycles

In temperate and arctic regions, seasonal changes in the length of daylight serve as proximate factors to regulate the reproduction of many animals. In the case of birds, however, photoperiodic reponses are modified by other environmental inhibitors and accelerators (p. 39). Lofts and Murton (1968) argue that, because most specialisation and adaptive radiation of the birds has occurred in tropical regions, where light plays comparatively little part in regulating annual cycles, it seems reasonable to suppose that the marked photoperiodic responses of north-temperate species may have evolved from mechanisms still existing in the tropics. They suggest that photoperiodically controlled cycles may have arisen from the autonomous cycles of tropical type by an increase in photosensitivity. Thus, the breeding cycle of the house sparrow *(Passer domesticus)* has evolved from the autonomous pattern displayed by equatorial species such as the red-billed dioch *(Quelea quelea).* On the other hand, the Andean sparrow *(Zonotrichia capensis)* has evolved from northern stock with a reproductive cycle controlled by photoperiod, but has secondarily invaded equatorial regions where, by convergence, it has evolved a cycle resembling that of *Q. quelea.*

Some species of birds have a high threshold of photosensitivity and require long daylengths before showing signs of gonadal recrudescence, while others with lower thresholds respond to much shorter periods. Lofts and Murton (1968) recognise six types of subarctic and temperate zone species of birds, depending on the photosensitive threshold levels, to which the various phases of their reproductive rhythm respond.

The rook *(Corvus frugilegus)* is an example of a bird with a relatively low level of photostimulation and an intolerance of higher levels of stimulation. Slight increases in day length early in the year are sufficient to bring the birds rapidly into breeding condition. Most of their egg-laying takes place in late March and early April, and is preceded by a rapid recrudescence of the gonads in late February and March (Marshall and Coombs, 1957), stimulated by the relatively short photoperiod of those months. Thus, in Britain, the breeding cycle of the rook is ultimately adjusted to coincide with the sharp peak in the numbers of earthworms available at the time when the young are being fed (Lockie, 1955). Moreover, the response can be accelerated or inhibited by temperature and other environmental variables.

A complete contrast is provided by species which do not have a refractory period in their reproductive cycle and can be artificially stimulated throughout the year. The turtle dove *(Streptopelia turtur)* migrates to central Africa for the winter and, therefore, to prevent unseasonal breeding, requires a safety device in the form of a refractory phase (Lofts, Murton and Westwood, 1967). On the other hand, the wood-pigeon *(Columba palumbus),* stock dove *(C. oenas),* and rock dove *(C. livia),* have physiological breeding seasons which can be accounted for in terms of differential photosensitivity. Gonadal recrudescence in the three species is stimulated at different times of year by the same seasonal changes in natural photoperiod. Stock doves feed mainly on the seeds of weeds, and can rear their young more successfully earlier in the season than can wood-pigeons, which tend to breed when cereal crops are available. (See discussion *in* Lofts and

Murton, 1968).

Bird numbers are affected by reproduction and migration. In a study of the bird population of an elm-maple forest in Illinois, Twomey (1945) found that numbers fluctuated continuously. Peaks in the spring and autumn corresponded with peaks in the numbers of insects: in summer, populations were comparatively stable, territorial limits had been established., mates selected and nests built. Of 105 species recorded, bob-whites *(Colinus virginianus)*, owls *(Strix varia)*, fleckers *(Colaptes auratus)*, hawks *(Butes* spp.), woodpeckers *(Centurus carolinus, Melanerpes erythrocephalus, Dryobates villosus* and *D. pubescens)*, jays *(Cyanocitta cristata)*, crows *(Corvus brachyrhynchos)*, tufted titmice *(Bacolophus bicolor)* etc. were perennial residents (15 spp.). The winter season is a one of minimum bird populations, made up of the tree-sparrow *(Spizella arborea)*, slate-coloured junco *(Junco hyemalis)*, brown creeper *(Certhia familiaris)*, red-tailed hawk *(Buteo borealis)*, etc. (13 spp.) and of perennial residents. Seasonal migrants comprised 50 species, aestival residents thirteen species, and incidental visitors fourteen species.

Although knowledge of the control of reproduction in birds is still somewhat fragmentary, it seems evident that the mechanisms, from the afferent hypothalamic level to the gonads, are generally similar. A great variety of reproductive patterns has arisen, primarily by adaptive evolution, with respect both to environmental and to internal information. Included are the feedback effects, used by the hypothalamus in its control of the gonadotropic activity of the anterior pituitary. The adaptive significance of these patterns lies primarily in the use of those sources of information that are most reliable in bringing the population into reproductive activity at a time when the probability of survival is greatest, both for the young and for the adults that produce them (Farner and Follett, 1966).

In temperate regions of the world, seasonal reproduction in mammals is under proximal control very largely by photoperiod. In some species, it is the most important factor influencing the sexual cycle; in others, the sexual cycle may be modified by nutrition, as well as by photoperiodism. A cool climate appears favourably to influence the initiation of sexual activity in sheep, while psychic factors, such as the presence of rams hastens the cycle (Hafez, 1952). Sexual periodicity, in general, is very much more pronounced in temperate regions than it is among animals living under tropical or sub-tropical conditions where the seasons are less marked. Photoperiod and temperature play an important part in synchronizing endogenous sexual cycles with the seasons of the year. The hypothalamo-hypophysieal mechanism brings the reproductive cycle into appropriate relation with seasonal environmental changes (Bullough, 1951; Frazer, 1959; Marshall, 1942), but the mechanisms of the inter-relationships between various environmental factors and their effects on gonadal development are not yet fully understood.

In addition to their effects on sexual cycles, seasonal environmental changes play an important part in the synchronization of physiological rhythms of wool and hair growth, coloration, and of the deposition of subcutaneous fat. Endogenous seasonal rhythms are enhanced by exogenous environmental factors, themselves causing changes in vegetation so that its nutritional effect supplements the other factors.

The influence of seasonal changes on the circadian rhythms of animals has not been studied in much detail.The data at present available shows how the effects of environmental factors may shift the times of the peaks of daily activity in all types of animals (Cloudsley-Thompson, 1966). To take but one example, Erkinaro (1961) recorded the activity of voles *(Microtus agrestis)* over long periods. He found that, in summer, they were active mostly around midnight but, in winter, they were more active during the day. In spring and autumn there were transitional stages, with two peaks of activity per 24 hours, one in the morning and another in the evening.

Seasonal cycles are very marked among the invertebrate fauna of deciduous woodlands, in many of which the life cycle is completed within a year, although the timing of the rhythm may vary in different species. Seasonal changes are also apparent in their behavioural responses (Cloudsley-Thompson, 1961a). A general account of the cycle of the seasons is given by Smith (1970).

The possession of a resting phase of 'diapause' enables many Arthropoda to survive the winter in a dormant state, characterised by enhanced resistance to cold and drought. Diapause is under hormonal control, and is usually induced by decreasing photoperiod. It has been the subject of much research (Beck, 1968; Lees, 1955). In general, the photoperiodic reaction is independent of the intensity and total energy of the light, provided that this exceeds a minimum threshold value. It is no coincidence that this threshold should exceed the intensity of moonlight.

Even long-lived invertebrates, whose life spans exceed one year, tend to show marked seasonal rhythms of activity. Although most of the Arthropoda are more active and more numerous in summer than in winter, a reverse effect is seen in slugs *(Arion hortensis* and *A. reticulatus).* These are far more active in winter than in summer. While their peaks of activity are quite definite, they vary from year to year according to the weather conditions (Barnes, 1944).

8 TAIGA

Coniferous forest or 'taiga' is able to exist in regions where the growing season is too short to support deciduous woodland. Because their leaves do not fall, but survive throughout the winter, coniferous trees are ready to begin photosynthesis without any delay and as soon as temperatures become favourable. Moreover, the coniferous type of fructification has the advantage that it is pollinated one year and dispersed the next, whereas deciduous trees have to complete the process within a single season. Length of growing season – that is, the number of months with temperatures above the threshold for growth – is clearly the most significant factor controlling forest types.

Deciduous trees pass through two distinct habits of life during the year – hygrophilous and xerophilous. They require rather long transitional periods in which to elaborate their transformations from one to the other. These are provided by the prolonged spring and autumn, which are found only in marine climates. The transition from winter to summer and back again is sudden and complete in continental climates which consequently are better suited to the coniferous, evergreen habit. Furthermore, coniferous trees require less precipitation – about 40 cm (15 in.)– than do deciduous trees. Deciduous forest may therefore be limited by deficient rainfall as well as by low temperatures (Miller, 1965).

The Siberian taiga is the largest forest in the world. It extends from the Pacific Ocean to the Ural Mountains and is some 5,800 km (3,600 miles) long by 1,300 km (800 miles) wide. In contrast to tropical rain-forest, in which a large number of different species are crowded together, the trees of the taiga belong to comparatively few species. They are mostly xerophilous in habit since they grow in soil which is physiologically dry and, in winter, are exposed to bitter, desiccating winds. Although most of the taiga remains in its primitive grandeur, it is threatened by destruction from tree felling which lowers the capacity of the soil to retain moisture, and increases run-off, especially on waterbeds.

Climate

Taiga is characteristic vegetation of cold or boreal and continental cool-temperate climates, where conditions are unfavourable for deciduous woodland. Cool-temperate climates differ from those of warm-temperate regions by the possession of a cold season which inhibits, or grossly retards, the active growth of plants. The severity of the cold season increases from west to east, away from marine influences, and there is a striking contrast between oceanic and continental regional climates. This is especially marked in the northern hemisphere where the two largest land masses of the earth are separated by the Atlantic and Pacific oceans. For their latitude their summer climates are the warmest in the world.

The prolonged cold of the continental winter is most extreme in January; that is, soon after the solstice. In the southern hemisphere, July is the coldest month of the year in boreal regions.

Cold climates lie within the sphere of influence of the westerly winds and are subject to controls which differ little from those of cool-temperate climates. There is the same eastward gradation from marine to continental type, making itself felt in the same way by an increase in annual and diurnal range of temperature, a decrease in rainfall and an increasing tendency to a summer maximum of precipitation. The land masses of the southern hemisphere do not extend sufficiently towards the pole to experience this type of climate which is, therefore, largely restricted to Eurasia and North America.

Where rainfall is adequate, coniferous forest reigns supreme but, where it is deficient, grassland takes its place. The climate of the taiga is similar to, but moister than, that of steppe (p. 108). The winters are cold and dry with long periods of frost. The summers tend to be short and comparatively hot. Rapid fluctuations of temperature occur, however, because the direction of the wind is irregular, one day importing warm air from nearer the equator, another blowing from the frozen poles.

Boreal lands have only very short periods that are entirely without frost. The growing season in the Mackenzie Valley of Canada varies from about 50 to 75 days, and a shift in the direction of the wind may bring frost, even in mid-summer. Precipitation is usually between 25 and 100 cm (10-40 in.) per annum and the mean temperature of the warmest month exceeds 10°C (50°F).

Vegetation

Rapid fluctuation in temperature imposes the necessity for great adaptability on the part of plants and animals whose survival depends upon this quality. The following main types of northern coniferous forest may be recognised.

(a) *Mixed coniferous forest.* This occupies most of the boreal forested parts of Eurasia and North America and is dominated by spruce *(Picea* spp.*)*, fir *(Abies* spp.*)*, pine *(Pinus* spp.*)* and larch *(Larix* spp.*)*

(b) *Open taiga.* Park-like taiga occurs towards the northern limit of the trees where the forest is sparse. It is characterised by widely-spaced trees; while associated vascular plants are equally scattered. Lichens, reindeer-mosses *(Cladonia* spp.*)* and Iceland mosses *(Cetraria* spp.*)* are typically aggregated into a pale but dense sward, several centimetres thick. In damp depressions and along the courses of rivers the forest is much denser, and tongues of timber even penetrate into the tundra.
(c) *Lake forest.* This lies in the eastern half of North America and is centred on the northern portions of the Great Lakes. The region is one of moderate precipitation – 60 to 115 cm (24-45 in.) and of considerable temperature extremes. The forest consists of associations dominated by white pine *(Pinus strobus),* Norway pine *(P. resinosa)* and hemlock *(Tsuga canadensis).* Various broad-leaved deciduous trees also occur in lake-forest which mingles with the deciduous summer forest to the south (Polunin, 1960).
(d) *Montane and sub-alpine forest.* Found in western North America, for example, this type of northern coniferous forest is due in part to local physiographic factors and, again, differs from true taiga.
(e) *Moist temperate coniferous forest.* Although some of this occurs in the same latitude as taiga, and forms part of the northern coniferous forest belt, it is quite distinct from true taiga and has been discussed in Chapter 7.

The taiga is a forest of mixed growth. In the north, larch, spruce and birch *(Betula* spp.) dominate but, in more southerly latitudes, these are interspersed with fir, pine and cedar *(Cedrus* spp.*).* Here and there are deciduous thickets of aspen *(Populus* spp.*)*, willow *(Salix* spp.*)* and rowan or mountain-ash *(Sorbus aucuparia).* The lianas and climbers characteristic of tropical rain-forest are replaced by the thin, wiry saplings of birches, alders and rowans which spindle up to heights out of all proportion to their girth in the race for sun and air – straight and close – set like the bars of a cage (Haviland, 1926). This deciduous growth usually occurs where the conifers have been destroyed by man, storm or fire.

The Siberian taiga is a low, rolling plain traversed by gentle ridges and shallow valleys which are often marshy and contain sluggish streams. These boggy places which, in the north, are perpetually frozen beneath the surface, are covered with mats of hoary sphagnum *(Sphagnum* spp.*)* and a few scattered spruce trees *(Picea obovata).* They correspond to the 'muskegs' or moss-bogs of the Canadian forests (p. 119).

Plants with juicy and fleshy seeds are common in the taiga. This may be due to the fact that dissemination of seeds by wind is impossible in the calm of the forest, and that dispersal by frugivorous birds and mammals has taken its place. Certainly seed-eating birds are numerous in comparison with insectivorous species (Haviland, 1926).

Forests subject to annual freezing are marked by discontinuity of species. They are populated by variously closely related species of spruces, firs, larches, birches and aspens. This belt is relatively new, much of it growing on sites that have been glaciated within the last 10,000 years. It is populated by species that survived the Pleistocene in cool and moist refuges, most of which were located in lower latitude highlands.

Except for eastern Siberia, where larches *(Larix* spp.*)* are more numerous, the circumpolar forest is elsewhere characterised by spruces *(Picea* spp.*)*. The principal species in North America are *P. glauca* and *P. mariana,* which range from Alaska to Newfoundland. *P. abies* is characteristic of western Europe while the closely related *P. obovata* extends to eastern Siberia. All are similar in appearance and in ecological habit. They are characterised by a shallow rooting system, often confined mainly to the humus layers of the soil. Their sharply conical crowns are well suited to bear winter snow and to shed it when the temperature rises or wind blows. Moreover, spruces are tolerant of undrained, acid soils, and can even grow in bogs.

Firs *(Abies* spp.*)* have a similar distribution to that of spruces but they are less tolerant of poor drainage and fire and better adapted to warmer, drier climates. While firs are common, upland species of the boreal forest of North America, they do not occur to any extent in the Palaearctic taiga. Instead, as already mentioned, the principal trees of eastern Siberia are larches *(Larix sibirica).*

In the boreal forests of North America, the genus *Larix* occurs only as tamarack *(L. laricina),* a species restricted to swamps by its relatively poor success in competition with other trees. Larches are deciduous and, for this reason, are well adapted to cold, dry climates: spruces and firs are better suited to cold, wet climates. Larches require open sites, however, and dominant positions for their crowns throughout life. They cannot compete successfully with spruces or firs under moister climatic conditions.

Birches *(Betula* spp.*)* and aspens *(Populus* spp.*)* are frequently associated with spruces and firs, but extend into drier regions. They are light-seeded, pioneer trees which can colonise, even after extensive burning, provided that their roots are undamaged. Their abundance is a direct measure of the severity and frequency of past forest fires. Like birches, aspens form a circumboreal complex of inter-grading populations that have migrated north from various centres of refuge after the ice-ages of the Pleistocene. *P. tremula* in Europe and Asia, *P. tremuloides* in North America and *P. suaveolens* in eastern Siberia, are the names given respectively to the Palaearctic and Nearctic portions of this complex (Spurr, 1964).

Fauna

Summer conditions present no special problems to the fauna of taiga. Temperatures are not extreme, and food and water are comparatively abundant. The stress comes in winter when famine and drought, produced by the intense cold, are important selective factors. Wind is not an important factor as it is in the tundra and steppe where it adds greatly to the severity of the winter.

Mammals. Except for a few rodents, such as voles, mammals remain hidden in the forest. They include bears *(Ursus* spp.*)* although these occur elsewhere, wolves *(Canis lupus),* gluttons *(Gulo gulo* and *G. luscus),* otters *(Lutra lutra* and *L. canadensis),* badgers *(Meles meles* and *Taxidea taxus),* stoats *(Mustela erminea),* sable *(Martes zibellina* and *M. americana),* lynxes *(Lynx lynx* and *L. canadensis),* elks *(Alces alces),* mooses *(Alces americana* and *A. gigas),* and the Siberian wild dog *(Cuon alpinus).* On the whole, taiga mammals are larger than the same forms on the tundra and steppe to north or south of them. Thus, forest reindeer *(Rangifer tarindus)* are larger than tundra deer, and forest wolves than steppe wolves. Tigers evidently evolved north of the Himalayas and later colonized India and S.E. Asia. The Siberian tiger *(Panthera tigris)* is larger than tropical forms and has a long, thick coat. The lemmings *(Lemmus* spp.*)*, however, provide exceptions to this generalisation. (See discussion of size in relation to latitude in Chapter 13).

The Canadian porcupine *(Erethizon dorsatum)* is largely arboreal, though it nests on the ground. Other arboreal mammals include squirrels *(Sciurus* spp.*)*. Apart from the lynxes, and the wild cat *(Felis silvestris),* the Felidae are little represented in the northern forests. Instead, there is an abundance of weasels, stoats, minks and polecats (Mustelidae) in both Palaearctic and Nearctic taiga. Many of these show seasonal change of colour and adopt white pelage in winter (p. 205).

Shelford and Olson (1935) found that most of the large ungulates and corresponding flesh eaters of the coniferous forest biome of North America range over the climax and subclimax stages, and often have a preference for the subclimax plant communities, though most of them are restricted to the area of the biome climaxes. These species exceed the limits of their biomes with a frequency about equal to that of the climax dominant plants. The data show the importance of considering animals in setting up biotic units. They indicate weaknesses in the concept of biomes based on plant data alone.

Birds. Bird life is comparatively plentiful in the taiga, but it is easily overlooked. Most species are migratory, but they do not winter so far south as do those of even more northern latitudes. Insectivorous species such as wagtails *(Montacilla* spp.*)*, flycatchers *(Musicapa* spp.*)*, swallows

(Hirundo rustica) and warblers are relatively scarce, but there is a disproportionately large number of thrushes *(Turdus* spp.*)*, finches and buntings (Fringillidae) and others that feed mainly on seeds and berries (p. 138). Crossbills *(Lixia leucoptera* and *L. pityopsittacus)* are completely adapted to life in coniferous forest. They live in the upper branches of the trees like little parrots, which they resemble somewhat in their gay plumage. Their beak is specially modified to split fir cones: the upper and lower mandibles are crossed so that the tips overlap (Fig. 20).

Fig. 20 Break of the crossbill *(Loxia leucoptera)*.

Grouse *(Lagopus lagopus)*, grosbeaks *(Pinicola enucleator)*, woodpeckers *(Picoides tridactylus, Dryocopus martius* etc.*)*, nutcrackers *(Nucifraga caryocatactes)* and jays *(Cractes infaustus)* are characteristic birds of the northern coniferous forests. A host of migrant species belonging to other groups joins them in the nesting season.

In summer, the bogs and streams of the tundra engender vast swarms of mosquitoes and support several species of wading birds. Sandpipers *(Tringa hypoleucos* and *T. glareola)* and oyster-catchers *(Haematopus ostralegus)* are found along the sandy beaches of the rivers, and the willow thickets and deeper woodland hold green and wood sandpipers *(Tringa ochropus* and *T. glareola)*. In Europe, the wood sandpiper often breeds on the ground but, in the taiga, like the green sandpiper and the

American spotted sandpiper *(T. macularia),* it nests in spruce trees, usually in old squirrels' dreys and fieldfares' nests. Three ducks of the taiga, the smew *(Mergus albellus),* goosander *(M. merganser),* and goldeneye *(Bucephala clangula)* likewise nest in trees at some height above the ground. The woodcock *(Scolopax rusticola)* spotted redshank *(Tringa erythropus),* great snipe *(Capella media),* common *(C. gallinago),* Jack snipe *(Lymnocryptes minimus)* are also characteristic of these swamps (Haviland, 1926).

Diurnal birds of prey such as eagles, falcons and buzzards are common, and the Siberian taiga contains many owls which are eminently suited to life in the dark forest. Common game birds include capercaillie *(Tetrao urogallus),* black grouse *(Lyrurus tetrix)* and hazel-hen *(Tetrastes bonasia).* Relatives of these birds are even more numerous in the taiga of North America.

In winter, the scales of the feet of willowgrouse *(Lagopus lagopus)* grow until the toes are surrounded by a horny fringe which supports the bird on powdery snow (Haviland, 1926). In the same way, desert-living lizards in various parts of the world have developed fringes to the toes which prevent them from sinking into loose sand (p. 104).

The reptiles of northern Europe have already been discussed in Chapters 6 and 7, and will not be mentioned again. Amphibia are represented by a few Bufonidae and Hylidae in North America and true frogs *(Rana* spp.*)* both there and in Eurasia.

Insects. The insect life of the taiga is rich. Coniferous trees are attacked by bark beetles (Scolytidae), among the most serious enemies the forester has to contend with, pine sawflies *(Lophyrus* spp.), *Lygaeonematus erichsonii* which is sometimes exceedingly destructive to larch, wood wasps *(Sirex gigas),* Geometridae, and processionary caterpillars *(Thaumetopoea* spp.*)*. The Diptera is the only other order of insects, apart from Coleoptera, Hymenoptera and Lepidoptera that has been able to adapt itself successfully to the climatic hardships of the taiga. The insect fauna of the Arctic is discussed in detail in Chapter 9, and many of the comments made there apply equally to taiga insects.

9 TUNDRA AND SNOWLANDS

Arctic climates can be divided into two types: (a) Tundra climates with a summer, however, short, above freezing. During this time the ground is free from snow for a sufficient period to permit the growth of typical tundra vegetation. (b) Polar or perpetual frost climates. In these, the growth of vegetation is impossible.

Tundra, a word of Finnish origin, meaning an open, forestless stretch of country, is applied to the huge tract of land lying north of the Arctic circle but which thaws in summer. In the southern hemisphere, the tapering of the land masses and their cessation at relatively low latitudes makes the development of true tundra impossible. Typical tundra is treeless but, as already mentioned, there are patches of stunted coniferous forests in valleys of rivers such as the Lena and Yenisei and along the coasts of northern Russia, which represent outliers of the taiga.

Climate

The inequality of the length of day and night reaches its maximum in Arctic and polar regions. At Lake Hazen (81°49' N) in northern Ellesmere Island, for instance, the sun is above the horizon for only 143 days each year (Downes, 1964). Diurnal range, therefore, has little meaning. Insolation is absent in midwinter and continuous at midsummer, however low the angle of the sun whose feeble rays have little power to melt the snow which absorbs most of their heat. In the tundra zone of the northern hemisphere, the temperature of the air does not rise above freezing-point until June and winter sets in by September. Any solid object may, however, be warmed so that black-bulb temperatures can exceed 38°C (100°F) although the air is freezing. At Bel-Sund, Spitzbergen, in July, when the ground was frozen below 30-35 cm and the air temperature at a height of 1 m was 4.7°C (40.5°F), the temperature just above low plants was 15.5°C (60°F). Maximum soil surface temperatures of about 50°C have been recorded at 73°N (Sørensen, 1941) and of 33°C at 82°N (Corbet, 1967).

Tundra climate is transitional between those of middle latitudes, usually boreal or cold continental, and that of the polar ice-caps. Typical tundra vegetation is found in a region whose equatorial and poleward boundaries are marked by the warmest month isotherms of 10°C (50°F) and 0°C (32°F) respectively. Over land, this climate is confined almost exclusively to the Northern Hemisphere since most of

Antarctica is polar.

Long, bitterly cold winters and short, cool summers are the rule. Usually only two to four months have average temperatures above freezing and killing frost can occur at any time. The diurnal temperature range tends to be small, even though the annual range is large. Indeed, the climate of the tundra is one of the bitterest on earth. In winter the temperature often drops to -57°C (-71°F). It is not, however, the severity and duration of the cold season, so much as the shortness and coolness of the summer that determines the character of the flora and fauna. Then the mid-day temperature may reach 21°C (70°F) and, although little rain falls, the splash and gurgle of water from the melting snow is heard everywhere. The boundary separating the tundra from the taiga to the south roughly coincides with the July isotherm of 10°C (50°F). Wherever the mean temperature of the hottest month exceeds this figure, tundra is replaced by forest (Haviland, 1926).

Although it may seem paradoxical, the swampy tundra has much in common with hot deserts. For, while deserts are physically dry, tundra is physiologically dry. Although rainfall is low, little evaporation takes place so there is plenty of water: but this is frozen and inaccessible to living organisms for most of the year. Again, in both regions, the vegetation has an extremely short flowering season. In deserts, this depends upon the brief rainfall; in tundra on the short summer thaw. Both tundra and desert are subject to strong, desiccating winds which bring the land under snow or sand respectively, and both have oases. In deserts, these are the dampest spots, in tundra the driest where the snow melts earliest and the soil is best drained (Cloudsley-Thompson, 1965d).

Polar climates have the distinction of producing the lowest mean annual, as well as the lowest summer, temperatures for any part of the earth because the rays of the sun are so oblique that they can never be genuinely effective, even during the long days of summer. In addition, much of the sun's energy is reflected by the snow and ice, or dissipated in melting the snow and evaporating the water thus produced, so that neither the land nor the air is warmed. Winters are bitterly cold, rivalling the severest in boreal regions. Consequently, annual ranges are large despite the cool summers. A conspicuous feature of polar climates is the occurrence of intense low-level temperature inversions.

Precipitation is meagre, often less than 25.5 cm (10 in.), but the amount of evaporation is also low, so that great permanent fields of snow and ice have accumulated on Greenland and the Antarctic continent.

The Antarctic, with its great land mass and high mountains, tends to have much colder winters and cooler summers than the Arctic. Summer temperatures remain consistently below the minimum necessary for the development of most higher plants and, consequently, only a handful of flowering species is found on the continent of Antarctica whereas some 400 species grow north of the Arctic circle. In addition, in the Arctic there are abundant mosses and lichens which do not die down in winter and, therefore, are permanently available as food for herbivorous animals.

Most animals and plants are found around ground level, where they receive maximum insolation during the day and are protected in winter by snow. Plants are seldom more than 20 cm high and, in winter, their lower leaves are sheltered between the ground and the overlying snow. Here live arthropods amongst the leaf litter and debris and, if the snow is thick enough, small mammals such as mice and weasels (Johnson, 1953). Birds, such as the Arctic redpoll *(Carduelis hornemanni)* may enter this zone briefly to breed (Cade, 1953). Where there is no overlying snow, the top of the zone is less defined because, with the sun near or below the horizon, there may be no eddy diffusion to provide a recognisable upper limit. In such cases, the zone is similar to, but narrower than, its counterpart in summer, and is inhabited by plants and by mammals (e.g. hares, wolves and musk oxen) which remain active during the winter (Corbet, 1972).

Vegetation

The vegetation of the green-grey tundra is exposed to extremely adverse circumstances which eliminate all but a few hardy species. A long period of frost is followed by an extremely short growing season. The seeding habit has been generally discarded in favour of vegetative reproduction and plants grow feverishly until nipped by the winter cold. When seeds are produced, they are extremely resistant to frost (see p. 201). although they require warmth to develop and do not germinate much below 20°C (68°F).

Much of the tundra overlies deep deposits of sphagnum which result, in part, from the failure of dead plants to decompose at low temperature. An area of wet tundra is traversed by the Hudson Bay railway which rests on frozen sphagnum to a depth of 6.5 m (20 ft.) near the southern edge of the tundra and to 2.6 m (8 ft.) near Churchill. The differences in depth depend upon topography (Clements and Shelford (1939).

Dwarf birches *(Betula nana)* and willows *(Salix arctica)* form a characteristic element of the vegetation of Arctic tundra. Ling *(Calluna vulgaris)* also forms a dwarf shrub, often about 50 cm high, but sometimes considerably more, with close-set, overlapping, scale-like leaves and open reddish-puce coloured flowers (Tansley, 1949). The low stature of the plants, and the closely tufted habit or rosette form which is common among them, gives protection from wind to their leafy shoots after the snow cover has been removed. Most plants are perennial and reproduce asexually: many possess rhizomes, bulbs or runners.

An outstanding feature of Arctic, as of Alpine vegetation, is the predominance of lichens and mosses which are especially resistant to rapid changes of temperature and humidity during the snow-free period. Where conditions are favourable locally, there may be a luxuriant growth of higher plants, even though they often do not flower. Presumably in such cases the seeds must have been transported from elsewhere by wind. Flowering plants often produce berries which are fed on by many birds and mammals, including the Alaskan grizzly *(Ursus horribilis).* Their leaves frequently have leathery, waxy or hairy surfaces which serve to reduce transpiration.

Fauna

Mammals

The mammals of the tundra have no special protection against the winter cold; nor can they migrate for long distances like birds do. The most striking species are the caribou *(Rangifer tarandus)*, lemmings *(Lemmus* spp.*)*, Arctic fox *(Alopex lagopus)*, arctic hare *(Lepus arcticus)*, wolf *(Canis lupus)* and stoat *(Mustela erminea).* The Siberian tundra also possesses a few species which are really forest forms, like the glutton or wolverine *(Gulo gulo)*, brown bear *(Ursus arctos)*, common fox *(Vulpes vulpes)* and some voles *(Microtus* spp.*)*

In summer the wild reindeer range up to the shores of the Arctic Ocean, frequenting chiefly the high ground which is comparatively free from mosquitoes. The fawns are dropped in May and the Samoyedes of the Yenisei say that at this time there is a 'Truce of God' between the deer and their enemies the wolves. Both species resort to the same parts of the tundra to breed and, for a short time, live there in harmony. As Haviland (1926) points out, however, the explanation is not that wolves have sentimental scruples but that, in summer, lemmings afford them abundant alternative food.

Caribou *(Rangifer tarandus)* are migratory and so do not over-graze the slow-growing lichens on which they feed, although they may move in enormous herds. In parts of Newfoundland, the rocks are worn away to a depth of half a metre by the thousands of hooves that have passed over them for countless years on migration. When domesticated reindeer were introduced into Alaska at the end of the last century their numbers increased to over half a million and then declined catastrophically. Their relatively sedentary habits had led to destruction of the food supplies essential for survival in winter. Complete recovery of the lichen requires at least 25 years, and its ecology is complicated

(Leopold and Darling, 1953). (p. 210).

Many Arctic mammals and birds have evolved relatively large feet which act as snow shoes. Ptarmigan and grouse have feathered feet with widened toes, while the pads of the hind legs of the snowshoe rabbit *(Lepus americanus)* are almost twice as long as those of an ordinary rabbit. Ungulates are sometimes unable to cross snow surfaces that will bear the weight of lighter predators such as wolves, lynxes and bob-cats. This hazard is countered, however, by making trails and areas where the snow is trampled hard. Canadian moose may sometimes be restricted to an area of less than 100 m radius. Ungulates frequently paw the snow to reach the underlying vegetation, and are often accompanied by ptarmigan which are thereby also enabled to feed.

The characteristic marine mammals of the poles, the walruses *(Odobenus rosmarus* and *O. divergens)* and seals, spend a considerable part of their lives on shore, especially during the breeding season. Some species of seal migrate to temperate regions in winter and thus avoid the intense cold, but others live permanently in the coldest regions and make breathing holes through the ice. In contrast to the comparative poverty of life on land, the polar seas are extremely rich in plankton which provides an abundance of food for fishes, birds and marine mammals.

Birds

Birds are the most conspicuous animals of the tundra in summer, but only the hardiest are able to winter in the north. Despite the tremendous distances they have to fly, most species arrive at the beginning of summer in a fat and healthy state. At first there is little food available and all their activities appear to be concentrated on establishing territories, courtship and mating. Later on, as the thaw sets in, mosquitoes appear in great numbers and their larvae are devoured not only by small waders and passerine birds, but by larger forms such as gulls, ducks and sandpipers. Most Arctic birds have a wide range of diet, stints and plovers even eating willow buds and golden plovers swallowing crowberries.

One of the reasons why plovers, in particular, are able to inhabit the frozen north is that, as in snipes too, both parents help to rear the brood. Moreover, not only do the young develop extremely rapidly, after hatching, thus shortening the period of dependence on their parents, but social tendencies are strongly marked, providing the protection of numbers. Arctic terns breed in colonies and will drive off enemies many times larger than themselves, diving at them and then swooping up to renew the attack.

In temperate climates, most birds begin to moult their summer dress as soon as their young are fledged. In the Arctic summer this process must be accelerated if the new plumage is to be ready for the autumn migration and feathers are shed so rapidly that the birds are often almost incapable of flight. In the far north, the ptarmigan has to moult three times between June and September – from winter to summer plumage, from summer to autumn and back to winter white again. This species does not migrate but digs tunnels in the snow where it finds both shelter and perhaps the occasional item of food.

In many parts of the Arctic, certain species of birds fail to breed when food supplies fail or weather conditions are unfavourable at the outset of the short season. Among these are eider duck and snowy-owls (Cloudsley-Thompson, 1965d).

Green plants, chiefly grass and Arctic willow *(Salix arctica)*, are the staple food of lemmings, grouse and geese. Other animals rely upon them to a lesser extent: reindeer browse on them and they provide cover for nesting birds (Haviland, 1926). Lemmings are the main food of foxes, wolves and birds of prey, mosquitoes and other insects of passerine birds and waders, which also feed on small aquatic animals.

According to Von Trautvetter (1848, cited by Haviland, 1926) the Siberian tundra can be sub-divided into four vegetational formations with their characeristic faunas, as follows:

(a) The general level of the drier tundra. These comprise vast tracts of higher ground whose only vegetation is lichen, interspersed with coarse grass. Flowering plants are almost entirely absent and the surface of the land is rough and broken. Mammals and birds are few – Arctic fox *(Alopex lagopus)* and lemmings *(Lemmus* spp.*)*, curlew-sandpiper *(Scolopax testacea)*, bartailed godwit *(Limosa lapponica)*, golden plover *(Charadrius apricarious)* and birds of prey.

(b) The flooded flats, including low-lying land along the banks and rivers consist of willow scrub or moss-bog. The former provides the only cover for the nests of willow-grouse *(Lagopus albus)*, Temminck's stint *(Calidris temminckii)*, white-fronted goose *(Anser albifrons)* and red-throated pipit *(Anthus cervinus)*. Moss-bog is flowerless except for a lousewort *(Pedicularis lanata)*. It is a favourite nesting ground of ducks and wading birds. In summer, vast swarms of mosquitoes provide an ample supply of food for the passerine birds which nest in the tundra during the summer.

(c) Slopes and declivities. These 'flower oases' are found on the south sides of stony hills, frequently above river valleys. The plants grow in mats and cushions, a few cm high, thickly sown with coloured blossoms.

Invertebrates

On account of the sparse flora, the Antarctic fauna is confined to birds and mammals that directly or indirectly depend upon the sea for their food. These include penguins and the skuas that prey on their young, seals and so on. Invertebrate animals are surprisingly few in number and, apart from parasites, are restricted to spring-tails, mites, tardigrades or 'water-bears', rotifers and a wingless chironomid fly that inhabits penguin rookeries. These are only active for a short period each year and for the rest of the time exist in a frozen state.

On contrast, Arctic regions have a comparatively rich fauna that is independent of the sea. But, of course, the numbers of species and of individuals decrease towards the pole as they do up the snow-covered slopes of mountains and relatively few are able to survive the most severe environmental conditions. Nevertheless, a remarkable fauna of earthworms, insects and arachnids inhabits regions of snow and ice. Their dark colours enable them to absorb the scanty warmth of the sun's heat upon which their metabolism depends. Dead insects, blown on to glaciers, are often to be seen in depressions where the warmth they have absorbed from the sun has melted for them an icy grave.

Earthworms of the family Enchytraeidae are often found frozen in ice. *Mesenchytraeus gelidus* can bore through hard packed snow, feeding on minute soil algae, but probably part of its life-history is spent in the soil.

The land and fresh-water snails of the Arctic are small in size and occur principally on the borders of springs. They are an unimportant element of the fauna. A number of orders of insects are represented of which flies, or Diptera, predominate as on high mountains. They are followed in numbers by the Hymenoptera – saw-flies, ants, bees and wasps – spring-tails (Collembola) and Lepidoptera. The composition of the fauna depends both upon available food and upon resistance to cold. Insects that feed on green plants do not penetrate so far to the north as do species that feed on lichen, plant remains or on animal food.

By splitting the rock with an ice axe I found a fauna consisting apparently only of spring-tails, mites and linyphiid spiders on a nunatak projecting through the ice cap Dranga Jokull in north-west Iceland. It seems likely that Collembola form a basis of food chains in regions of snow and ice, as Thysanura do in deserts. One species, *Isotoma nivalis,* the 'snow-flea' of Spitzbergen, like the larger Alpine *I. saltans,* forms great black masses sometimes extending over an area of about a square foot or more.

Mosquitoes, black-flies and other Diptera form a veritable plague during the short Arctic summer. They have in common the possession of an aquatic larval stage. Undoubtedly the commonest ground-living animals of the Arctic are wolf-spiders (Lycosidae) and, in Iceland, the harvest-spider *(Mitopus morio)* (Fig. 21) (Cloudsley-Thompson, 1965d).

Dominant insects of the tundra are Diptera, Hymenoptera, Coleoptera, Lepidoptera, Collembola, Ephemeroptara, Trichoptera, Odonata, Hemiptera and Orthoptera. Orthoptera and Hemiptera are

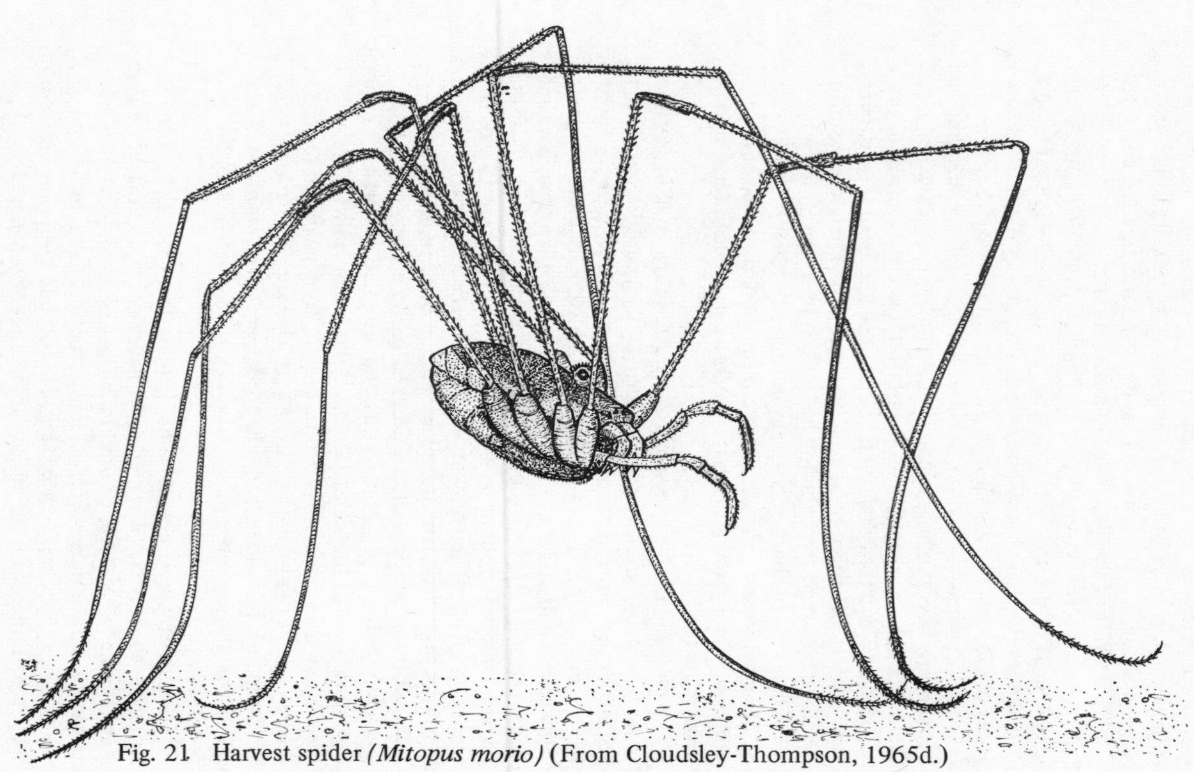

Fig. 21 Harvest spider *(Mitopus morio)* (From Cloudsley-Thompson, 1965d.)

confined to the south: orders with aquatic larvae occur a little further northwards, but in the far north only the endopterygote orders and Collembola. The Diptera are represented mainly by Tipulidae, Chironomidae and other Nematocera, and Anthomyzidae, Heteromyzidae and Syrphidae among the Brachycera. Caribou *(Rangifer tarandus)* are tormented by nostril flies and warble flies (Oestridae). The former parasitise the nasal passages and sinuses; the latter form swellings beneath the skin of the back.

Beyond the northern limits of the taiga there are no wasps or bees, except for Bombidae and a few species of Ichneummoidea. An outstanding feature of the tundra is the predominance of day-flying Lepidoptera, except in Iceland where the order is poorly represented. Genera such as *Colias, Argynnis, Chrysophanes* and *Lycaena* are among the most northerly ranging of butterflies, *Anarta, Plusia, Cidaria, Hyphoraia* and *Penthina* among moths (Haviland, 1926).

Most tundra insects require two summers or more to complete their development and tend to be somewhat smaller than their congeners further south. Melanic forms are common and there is frequently a thick clothing of scales or hairs which is specially noticeable in bumble-bees.

K. Zetterstedt (1840, cited by Haviland, 1926) listed the insects of Lapland according to the number of species per order, viz.:

Diptera	1245 spp.	(35%)
Coleoptera	1001 spp.	(28%)
Lepidoptera	499 spp.	(14%)
Hymenoptera	426 spp.	(12%)
Hemiptera	232 spp.	(6%)
'Neuroptera'	123 spp.	(3%)
Orthoptera	14 spp.	(0.3%)

while G. Jakobson (1899, cited by Haviland, 1926) arranged the orders of insects inhabiting the unforested region of Novaya Zemlya as follows: Diptera, Hymenoptera, Coleoptera, Lepidoptera, Collembola, 'Neuroptera', Hemiptera, Orthoptera. Flies are dominant in the north, and only in Iceland are beetles and bees equally plentiful. ('Neuroptera' was held to include Trichoptera, Ephemeroptera and Odonata.)

The Tipulidae and Chironomidae are the dominant families of Nematocera, the Anthomyzidae, Heteromyzidae and Syrphidae among the Brachycera (Jakobson, *loc.cit.*). Tundra beetles belong chiefly to families that are carnivorous or feed on carrion such as Carabidae, Dytiscidae and Staphylinidae: vegetable feeders, such as weevils (Curculionidae) are poorly represented.

The composition of the Arctic insect fauna is very different from that of temperate regions, and the number of species very much lower. In the low Arctic the number may be reduced to about 5 per cent and, on Queen Elizabeth Island, it is only about 1 per cent of the number found in comparable temperate areas. Many of the endemic forms are adapted to special features of the environment, especially the low temperature and short season. Others show a loss of specialised features such as colour or pattern.

The Arctic environment is not inherently simple according to Downes (1964), nor does it forbid a greater diversity. It seems to lie beyond the range of physiological tolerance of all but a very few of the forms of temperate origin. The greatest continuity of evolutionary history has occurred in the tropics (Fisher, 1960), to which are adapted most forms of life. Given time, however, there is no apparent reason why a greater diversity of Arctic forms should not develop (Downes, 1964).

Seasonal rhythms

Animals that live in Arctic regions must be able to survive the change from the cold and darkness of winter to the warmth and light of summer and vice-versa. Some of them pass the winter sheltering underground; others remain in the open, taking cover only during the worst storms. Nearly all the birds migrate to warmer climes before the winter starts.

During the winter, the ground is covered with a deep layer of snow. This provides cover or insulation for animals, such as lemmings *(Lemmus* spp.*)* which live beneath it. These neither store food in summer nor hibernate. In winter they dig long tunnels in the snow just above ground from which they search for plants and especially grass roots. They still keep their burrows in the earth, but cannot enlarge these because the soil is frozen solid.

Arctic hares *(Lepus arcticus)* live above ground and continue feeding throughout the winter, usually in places where the snow has blown away and left the plants exposed. The Arctic fox likewise feeds throughout the winter, but has difficulty in finding enough to eat. It begins to store food – birds, eggs, lemmings and so on – before the winter begins, hiding it in a cache underneath rocks, or covering it with gravel.

Most species produce their young at the beginning of the summer when the weather is warmer and food plentiful. The female musk ox bears a single calf at the end of April. It can follow its mother an hour after birth, but she feeds it for a year or more. Not many birds remain in the Arctic during the winter. Ptarmigan *(Lagopus mutus)* and snowy owls *(Nyctea scandiaca)* are exceptional in this respect. Ptarmigan feed on moss and lichen as well as on shoots of heath and ling, and berries when these are available. Insects form a large part of the diet of the chicks. Snowy owls are predators on lemmings and other small mammals, waterfowl and even fish, though their heavily muffled feet seem most unsuited for fishing. This large species hunts both by day and by night. As soon as spring begins, however, birds arrive in hundreds of thousands to feed and breed in the continuous light of the Arctic summer. Many of them are sea-birds and aquatic species which find their nesting places on cliffs or on islands along the coasts.

Population cycles

Few animal populations remain constant in size for very long. In addition to seasonal cycles in numbers, cyclic variation in numbers over periods of several years is seen especially among the mammals and birds of the Arctic and cool temperate regions of the world. Two or three main cycles seem to occur. First, there is a three-year or four-year cycle in the numbers of various mammals whose food is based on the lemmings *(Lemmus* and *Dicrostonyx* spp.*)*. This is seen, for example, in the white or Arctic fox *(Alopex lagopus),* and in the snowy owl *(Nyctea nyctea)* which is often forced by starvation to migrate hundreds of miles to the south (Chitty, 1950). Secondly, there is a four-year cycle in the animals of the belt of open forest lying between the tundra and the taiga. This is based on voles *(Microtus* spp.*)*. Thirdly, there is a ten-year cycle in populations of the snow-shoe rabbit *(Lepus arcticus)* and other animals of the northern forest regions of North America.

The extensive literature on four-year cycles has been reviewed by Elton (1942) and others, and the significance of the problem is discussed by Lack (1954) who suggests that the dominant rodent interacts with its vegetable food to produce a predator-prey oscillation. When the numbers of rodents decline, predatory forms themselves also decrease, thereby allowing an increase in gallinaceous birds. The regularity of the cycles may be because the basic predator-prey oscillations are less disturbed by other factors than in more complex biomes. (See discussion *in* Cloudsley-Thompson, 1961a). The phenomenon is, however, little understood. For instance, at least twelve different explanations have been published to explain population cycles in the ruffed grouse *(Bonasia umbellus)* (Bump *et al.*, 1947).

Cyclic fluctuations in numbers have two main advantages to a species; they permit correlation of numbers with changes in environmental conditions, and they allow evolutionary change to be more rapid than it would be in a species with constant numbers. This is because while the numbers of a population are increasing, selection is weak since a larger proportion of individuals are able to survive than when numbers are constant: consequently the population becomes more than normally variable. When it reaches its maximum, however, this variable opulation is subjected to natural selection. Any variations that happen to be advantageous are then selected into the genotype. Thus a biological advantage accrues to species whose numbers tend to fluctuate cyclically, in addition to any incidental increase in distribution caused by emigration.

Unless cyclical fluctuations in animal populations have an ultimate biological significance, it is difficult to understand how the phenomenon can persist in nature. The ultimate factors already discussed may, however, be quite different from the direct causal, or proximal, factors which may have been evolved as a result of natural selection and have a more direct influence upon individual animals or groups (Koskimies, 1955). Whereas, in the case of Arctic forms, food may be the ultimate factor that determines the general level of population numbers, Siivonen and Koskimies (1955) have suggested that the regularity of the fluctuations may be governed through an adaptive proximal response to the lunar cycle.

All attempts have failed to find a cause for these rhythms by correlating them with weather conditions, sun-spots and other cosmic factors. Siivonen and Koskimies, however, have described a theoretical mechanism based on the lunar cycle, which makes it possible to explain both the ten-year fluctuation and the three- and four-year fluctuation on a common basis. The length of the lunar month is 29.53 days. As the total of the twelve whole lunar months in one year (365.25 days) is 354.4 days or 10.9 days less than a year, the lunar cycle runs each year 10.9 days ahead of that of the preceding year. Thus each moon occurs 10.9 days earlier than its equivalent in the previous year. After a certain number of years, a given phase of the moon must return to the same data as at the beginning of the period. Owing to the length of the lunar cycle, a given phase returns to within six days of the same date at intervals of three to four years and to within one day every 9.6 years. Needless to say, the above explanation has not been generally accepted even though no better hypothesis has yet been proposed.

Adaptations of Arctic animals

Migration. Whereas the majority of insects pass the winter in a resting stage and birds migrate to warmer climes, Arctic mammals do neither. Hibernation does not take place – probably because the short, cold summer does not allow sufficient accumulation of the necessary food reserves. Also, the large amount of snow partly removes the need for hibernation because it provides an insulating layer in which small animals can burrow. At the same time it enables them to browse on leaves and twigs that would otherwise be beyond their reach.

With regard to migration, it is true that deer move southwards towards the forest and wolves follow them, but this provides but a slight amelioration of the cruel conditions in which they have to survive. In any case, smaller forms such as lemmings remain in their summer haunts, feeding on herbage buried under the snow (Cloudsley-Thompson,

1965d).

Lemmings from Greenland migrate over the frozen ocean to the mainland from islands 50 km distant. When the ice breaks, the little creatures run backwards and forwards along the banks of streams and rivers, looking for a smooth place with a slow current at which to cross. Having found one, they at once jump in and swim fast to the other side where they give themselves a good shake as a dog would, and then continue their journey as if nothing had happened (J. Rae, 1852, cited by Haviland, 1926).

Lemmings move erratically and no definite path is chosen, but the general direction is determined partly by the nature of the ground traversed. Water is no obstacle but the legend of self-destruction may have arisen out of incidents such as that related by R. Collett in 1868, when a steamer in Trondjem Fiord passed for fifteen minutes through a pack of swimming lemmings. Migrating hares and rabbits are also known to swim rivers and, in 1867, a swarm of squirrels is reported to have invaded Tapilah in the Urals, swimming the Tchossoveia River and climbing up the oars of boats which crossed its track (Haviland, 1926). Lemmings can swim across fiords more than 2.5 km in width. It is therefore not surprising that they should sometimes be swept away by the tide.

Size. The relationship between size and surface area (p. 194) is particularly important in the heat-relations of homeothermic animals living in the colder regions of the world. Reduction of the body surface is an important means of heat conservation and is achieved by the development of a compact form, with reduction of the appendages, or by an increase in size. Thus the musk ox *(Ovibos moschatus)* has legs so short that it stands only 1 m high though it is 2.5 m long. Its neck is thick, its tail very short and its ears are hidden in its furry coat (Fig.22).

Fig 22 Musk ox *(Ovibos moschatus)*

Arctic mammals conserve heat with their thick hair. Small species, such as foxes, hares and lemmings, have a silky fur and woolly undercoat, while reindeer and musk oxen have hairs thicker at the tip than at the root so that they form an almost airtight coat. In addition, a thick layer of subcutaneous fat provides insulation and acts as a food reserve. Carnivorous mammals nevertheless fare miserably in the Arctic winter and drag out a bare existence. Bears, wolves and foxes suffer extremities of famine; they become very emaciated and their stomachs are empty for long periods.

Colour. Most polar mammals and birds are either always white in colour, like the polar bear *(Thalarctos maritimus)*, or else show seasonal changes and become white in winter. Thus both predators and prey become less visible on the snowy landscape. In this, they contrast with poikilothermic species which are almost all dark so that the greatest amount of heat is absorbed during the brief season of their activity. Ermine is the name given to the Arctic stoat *(Mustela erminea)* when it has adopted its white winter coat. Even when ptarmigan *(Lagopus mutus)* are moulting so that only some of their feathers are white, they are still inconspicuous among the spring and autumn snow patches (Cloudsley-Thompson, 1965d). (See p. 205).

Supercooling. Physiological adaptations lie somewhat beyond the scope of the present volume. Nevertheless, the effects of cold on poikilotherms should be mentioned. These are manifold, but a primary distinction can be drawn on the basis of whether, or not, freezing results (Salt, 1961; 1964; 1969). Considerable research has been carried out on the physiology of survival of Arctic insects at low temperatures. In some cases, insects are able to escape the lethal effects of freezing by supercooling. The body temperature can drop as low as -20°C before crystallisation occurs. This may be effected by synthesising glycerol and other cryoprotective compounds before hibernation. Other factors, however, and especially the presence of nucleating agents in the alimentary canal would, as Salt has shown, appear to be of greater significance. The problem is a complex one, and clearly has no simple answer (Cloudsley-Thompson, 1970b; 1973).

Pleistocene changes

During the Pleistocene period, the polar ice caps advanced and retreated on a number of separate occasions. Several theories have been proposed to account for this. Of the more important, one is based on astronomical considerations, another on rhythmical variations in the sun's radiation. The ice ages had a profound influence on the evolution and ecology of man who, if he had not discovered the use of fire, certainly could not have survived in parts of Europe where he is known to have lived during the severe climate of the last glaciation.

Man's use of fire dates from the appearance of Pekin Man *(Homo erectus)* at Choukoutien, about 400,000 years ago, at the time of the second glaciation (Mindel 11) in Europe. Before this time, the ancestors of man had passed a long period of evolutionary development in warm, sunny regions (Weiner, 1971). The vicissitudes of the Pleistocene may therefore have played an important role in shaping human destiny.

10 MOUNTAINS

On account of the low temperatures and short summers, trees do not grow at high altitudes. The region above the timber line is known as the 'Alpine zone' in all parts of the world. It is, in many ways, similar to the Arctic regions discussed in the previous chapter. There are certain differences, however. At high altitudes, atmospheric pressure is reduced. Photoperiod is related to latitude rather than to altitude and, at high elevations, the atmosphere is thin. Consequently, insolation is powerful on mountain tops, in contrast to Arctic snowlands where the rays of the sun are strongly filtered as they pass obliquely through the atmosphere.

Climate

By causing the ascent of horizontally moving air, mountains engender increased precipitation and, in arid lands, may even result in altitudinal oases (p.19). Most of the water vapour in the atmosphere is, however, concentrated in its lower layers. Consequently, the air on the tops of high mountains is excessively dry, a feature enhanced by low pressure which allows evaporation to take place very rapidly. The absence of moisture makes the air clearer than it is at sea level and this, in turn, permits the passage of more ultra-violet rays than does air at lower altitudes. In addition, low-density air does not absorb much heat from the sun. But the ground does, so there is a wide divergence, during the day, between ground and air temperatures.

Winds reach high speeds on mountain tops because frictional drag with the earth's surface is reduced. Higher wind speeds increase evaporation and lead to even lower temperatures than would otherwise be found.

Vegetational differences due to topography are usually correlated with variations in moisture, especially where water is deficient. For this reason, the physiographic effect of aspect is important. In the northern hemisphere, slopes that face the north tend to be moister than slopes facing south where insolation is greater, even though precipitation may be the same on both. For this reason there may be entirely different floras on the two sides of a deep valley or steep mountain. In a similar way, moss campion *(Silene acaulis)* flowers only on the top and south facing sides of its domed tussocks in Spitzbergen where it can constitute a reliable compass. The effect is here though to result from differing light intensities (Polunin, 1960).

The steepness of the slope itself determines the stability of the surface, the retention of water, the degree of insolation and the amount of soil which accumulates. Steep slopes are eroded more rapidly than gentle slopes; the material sliding away as 'talus' to be deposited further down the mountain side.

Vegetation

The number of floral and faunal zones on a high mountain is greatest in the tropics, since its base may be in hot tropical lowland and its top above the permanent snowline. Every such mountain shows four or five belts or zones, whose width is proportional to the steepness of the slope.
(a) *Tropical belt.* The climate is hot, and dry or moist, according to circumstances. The vegetation tends to be evergreen, unless there is a pronounced dry season, when it is deciduous. This zone ends at a height about 900 m (3,000 ft.) above sea-level.
(b) *Warm temperate belt.* Here the difference between summer and winter begins to be marked and the nights may be cool. Vegetation is chiefly evergreen unless the dry season is unusually prolonged.
(c) *Cool temperate belt.* This often coincides with the cloud level. The winter season is well marked, and deciduous trees lose their leaves at this time. The upper limit of this belt coincides with the upper 'tree line'.
(d) *Alpine belt.* Characterised by grassy slopes with tundra vegetation of short, matted sedges, grasses and shrubs, and an abundance of wild flowers: higher up, the grass is replaced by mosses and lichens.
(e) *Arctic or 'nival' belt.* The lower limit is near the permanent snow-line which, in the tropics, lies somewhere near an elevation of 4,570 m (15,000 ft.) (Gadow, 1913). Only a few hardy flowering plants are found, with lichens encrusting bare rock surfaces.

These belts are comparable with the latitudinal vegetational belts described in preceding chapters. Mountains situated in high latitudes have fewer vegetational belts and those located in Arctic regions may lie entirely within the snow-line. Whatever the vegetation zone in which the base of a mountain stands, the zone of maximum precipitation is nearly always forest – deciduous below, but passing upwards into coniferous. Between the tree-line and the snow-line, in the region of steadily decreasing precipitation, stretches a zone of grassland – narrow in temperate regions where the snows are not far above the limit of the forest, but wide in the tropics and becoming increasingly dry towards its upper limit.

The altitude of the snow-line depends upon the latitude of the mountain range, the direction of slope and the nature of the climate. The highest forests reach 4,600 m (15,000 ft.) in Tibet: the Alpine zone begins at 3,600 m (11,800 ft.) on the south slope of the Himalayas, and at 2,800 m (9,200 ft.) in the Colombian Andes. Even at the same latitude (48°N), the timber-line varies from 3,500 m (11,500 ft.) in the Rocky Mountains to 2,000 m (6,600 ft.) on Mt Rainier. In Norway at 74°N it is only 260 m (850 ft.) above sea level: in New Zealand, at 39°S, it lies at about 2,000 m (6,600 ft.). The Alpine fauna ranges downward into the forest zone along glaciers which may descend far below the normal timber line (Hesse, Allee and Schmidt, 1951).

In the coniferous forests of the mountains of central Europe (Plate 4a), the most important species is the white fir *(Abies alba)*. It is the major tree species of the Pyrenees but, in the mountains of the Caucasus and the Black Sea region of Turkey, it is replaced by *A. nordmanniana*. The silver fir *(A. pindrow)* occurs in the Himalayas, *A. cilicia* in the mountains of Asia Minor, *A. pindow* in the mountains of northern Indo-China, *A. guatemalensis* and *A. religiosa* in the mountains of Guatemala and Mexico.

South of the taiga, outliers of birch *(Betula* spp.*)*occur at lower altitudes in the major mountain ranges of eastern North America, and in the Pyrenees, Himalayas and Korean mountains. They contain species less tolerant of cold than are the birches of the boreal forests: some colonise mountain slopes, others are riverain. They have only a limited distribution in the mountains of western North America, however, whereas aspen *(Populus tremuloides)* extends southward at high elevations down the Rockies and Sierra Nevada to the mountains of northern Mexico. In the Nearctic region, *Populus tremuloides* and *P. grandidentata* are the principal species. In Eurasia, outliers of *P. tremula* occur in the Himalayas, China, Korea and Japan, while *P. alba* is native to moister sites from central Europe through the Balkans and eastwards into Iran (Spurr, 1964).

Pines of temperate climates are found in the mountains of India, Pakistan, China, Korea and Japan. In the Himalayas region, chir pine *(Pinus longifolia)* grows in the hot and dry lower elevations while the blue pine *(P. excelsa)* is the principal species at higher altitudes. In China, *P. massonia* is the chief species while, in Japan, both black pine *(P. thunbergia)* and red pine *(P. densiflora)* are important.

In the mountains of the western range of North America, open types of ponderose pine *(Pinus ponderosa)*, Douglas fir *(Pseudotsuga menziesii)*, aspen *(Populus tremuloides)*, spruce *(Picea* spp.*)* and fir *(Abies* spp.*)* support stands of grasses, herbs and shrubs; while grass and sedge meadows occur in non-forested areas of the mountains. Above the timberline is found Alpine tundra.

Sub-alpine forest, which clothes the upper slopes of high mountains throughout the extent of the Rockies, forms a crisp and vigorous habitat with a distinctive flora and fauna. Rainfall varies from fairly ample to less than occurs in the boreal forest at lower elevation (p. 129). Sub-alpine forest is cool and growth rates generally low. Dominant trees include *Picea engelmanii* and *Abies lasiocarpa*. The timber-line may be delineated by climate, fire or edaphic factors. Snow slides or rock falls may lay bare stretches through the forest on steep slopes.

On the principal mountains of East Africa, three distinct vegetational regions can be recognised (Coe, 1967). These belts are often subdivided into a number of zones which are not all developed on every mountain. Montane forest forms a belt often divided into a lower montane rain-forest zone, an intermediate zone of bamboo *(Arundinaria alpina)*, and an upper *Hagenia-Hypericum* zone. The moorland belt is usually dominated by tree-heathers of the genera *Erica* and *Philippia* but with broad-leaved trees such as *Rapanea* spp. and *Hypericum leucoptychodes* in the lower part. The uppermost Afro-alpine belt around the peaks has a vegetation that varies from moist forest or arborescent species of *Senecio* and *Lobelia* on Ruwenzori, to open desert-like grass communities on Mt. Meru (Livingstone, 1967). Only three of the highest mountains, Kilimanjaro, Ruwenzori and Mt. Kenya, have permanent snow throughout the year. Like glaciers elsewhere in the world, those on the mountains of East Africa are steadily retreating and have disintegrated considerably, even during the present century.

Fauna

The fauna of high mountains is composed partly of animals peculiar to it, but largely of species that range into milder climates. Few birds or mammals exist for long above 6,100 m (20,000 ft.) in the Himalayas. The snow partridge *(Lerwa lerwa)* feeds and nests almost to this height, but other vertebrates appear to be visitors only. They include scavengers such as choughs *(Coracia graculus)*, lammergeiers *(Gypaetus barbatus)* and other vultures that follow man and domestic animals – yaks and sheep – wherever they go. The wild sheep or bharal *(Pseudois nahura)* is preyed on by snow leopards *(Panthera uncia)* which, together with wolves and foxes, may pursue it in their wanderings to very high altitudes. The small Tibetan weasel *(Mustela altaica)* feeds on the eggs of the snow partridge and occurs wherever these birds are found, while mice of various kinds sometimes ascend to considerable heights (Cloudsley-Thompson, 1965d).

Mammals

Most truly Alpine mammals are herbivorous. With the exception of the snow leopard, which tends to remain at high elevations in central Asia, predaceous species usually only range above the snow-line when there is sufficient food to tempt them. Some of the mustelids of North

America, such as the wolverine *(Gulo gulo)*, range southward in the Rocky Mountains, where they are confined to the Alpine zone.

A wide variety of sheep, goats and antelopes inhabit mountains; but the yak *(Bos grunniens)* is the only bovid. Wild sheep and goats are very agile and capable of extraordinary leaps. In South America, they are replaced by llamas, vicunas, alpacas and guanacos. Of the goat-antelopes, the chamois *(Rupicapra rupicapra)* is found in the Alps, Pyrenees and Caucasus, the serows *(Capricornis* spp.*)*, gorals *(Nemorhaedus* spp.*)*, takins *(Budorcas* spp.*)*, and orongo or chiru *(Pantholops hodgsoni)* in North America. Ungulates of mountains tend to show a restriction in their breeding seasons, even at comparatively low elevations (Fraser, 1968).

The smaller rodents include the snow vole *(Microtus nivalis)* and marmots *(Marmota marmota)* of the Alps, chinchillas *(Chinchilla laniger* and *Lagidium peruanum)* in the Andes. These are abundant at 3,000 m (9,800 ft.) and range to 5,000 m (16,400 ft.) (Hesse, Allee and Schmidt, 1951).

Birds

Insectivorous birds are scarce in Alpine regions on account of the scarcity of their normal food. Ground living species are, perhaps, best represented. These include the Himalayan snow partridge and the white-tailed ptarmigan *(Lagopus leucurus)* of the Rocky Mountains. Other birds confined to the heights include the Eurasian wall creeper *(Trichodroma muraria)* and snow finch *(Montifringilla nivalis)* of the Alps and Pyrenees. Carnivorous scavengers are the lammergeier of the Old World and the giant Andean condor *(Vultur gryphus)*. Golden eagles *(Aquila chrysaetus)* range over long distances for food, and frequently fly up into the Alpine and nival zones.

Steep slopes, characteristic of mountainous environments, increase the erosion which, in any case, proceeds rapidly owing to strong winds, heavy rains and extreme diurnal temperature fluctuations. In consequence, rock surfaces are frequently laid bare and this favours hyraxes and birds such as lammergeiers, buzzards *(Buteo* spp.*)* and African white-necked ravens *(Corvultur albicollis)* which nest in rocky situations.

Many accentors, small, sparrow-like birds related to thrushes, live in the high mountains of Eurasia. In summer they feed on insects but, in winter, they migrate to lower altitudes and eat seeds and berries. The Himalayan accentor *(Prunella himalayana)* breeds in Tibet at altitudes above 5,200 m (17,000 ft.). One genus of humming-birds *(Oreotrochilus)* is confined to altitudes of 3,000–4,900 m (10,000–16,000 ft.) in the Andes. *O. chimborazo* is found only on Mt. Chimborazo and various races of other species are each restricted to their own isolated peaks. Flowers are rare in the high Andes so these little birds do not get much nectar but feed, instead, mostly on insects. Their nests, made of fern fronds, feathers and cobwebs, are built either on the edges of cliffs where they obtain maximum advantage from the sun, or deep inside caves, sheltered from the wind and cold. Alpine choughs *(Coracia graculus)* which live in the mountains of Europe, Asia and North Africa, have been seen at 8,200 m (27,000 ft.) on Mount Everest and occupy the highest habitat on earth (Matthews and Carrington, 1970). In winter, however, they descend to the valleys.

Reptiles

Reptiles are not common at high altitudes but the Andean lizard *(Liolaemus multiformis)* reaches the snow-line in Peru, and *Lacerta agilis* a height of 4,100 m (12,500 ft.) in the Caucasus. *L. multiformis* has been seen emerging from its burrow when the air temperature was -5°C (23°F). It can warm its body to above 30°C (86°F) by the absorption of solar radiation, even though nearby shade temperatures are below freezing (Pearson, 1954). The viper *(Vipera aspis)* lives at altitudes up to 2,960 m (9,700 ft.) in the Pyrenees, *V. berus* up to 2,800 m (9,300 ft.) in the Alps, and garter snakes *(Thammophis sirtalis)* to 3,000 m (10,000 ft.) in the Rocky Mountains. Other examples are cited by Hock (1964b). All species found above 3,00 m (10,000 ft.) are ovoviviparous, as are Arctic reptiles (p. 113). Lizards of the genus *Phrynocephalus* in central Asia are oviparous at low altitudes and ovoviviparous at higher levels. The same is true of the iguanid genus *Sceloporus* in Mexico. Many mountain-dwelling species of lizards have characteristically short limbs.

The highland chemeleons *(Chemaeleo* spp.*)* of central Africa, which range up to 3,200 m (10,500 ft.) or more, are ovoviviparous, as are skinks *(Mabuya* spp.*)* which extend up to 4,000 m (13,200 ft.). *Algyroides alleni* is the only resident reptile in the Alpine zone of Mt. Kenya, where it is found among tussocks and under stones around the tarns, feeding largely on beetles and their larvae. Spellerberg (1972a) has demonstrated altitudinal zonation among Australian lizards of the *Sphenomorphus quoyi* species complex, correlated with tolerance of low temperature. The smallest species *(S. koscuiskoi)* occurs at highest altitudes; the largest *(S. quoyi)* is restricted to lower levels (Spellerberg, 1972b). Their preferred temperatures are similar, however, and maintained by shuttling between sunlight and shade. At higher elevations, more time is spent basking in the sun than in shaded areas (Spellerberg, 1972c).

Amphibians are found at high altitudes only where there is plenty of moisture. The altitude record is held by the green toad *(Bufo viridis)* which is found in the Himalayas at altitudes up to 5,000 m (16,400 ft.).

The length of larval life is shortened in species from mountains where breeding pools are free from ice only for short periods (Hesse, Allee and Schmidt, 1951). Amphibians do not occur much above the forest belt on African mountains, and no species seems to have become so specialised for life at high altitudes that it is not found also at lower levels (Cloudsley-Thompson, 1969).

Invertebrates

The Alpine fauna is, in many ways, similar to that of the tundra and Arctic. Invertebrate animals are dark coloured, and black earthworms have been found at 3,840 m (12,500 ft.) in Kashmir and in the snows of Kilimanjaro. Snails are limited to a few widespread species that tolerate drought and cold. Species of *Vitrina* range up to 4,400 m (14,500 ft.) on Kilimanjaro and rock snails of the genera *Clausila* and *Camphylaea* are abundant. Reduction in the size of the shell is a frequent occurrence at high altitudes. Reduced temperature and lack of food are the chief factors limiting the upward distribution of invertebrates. The snow, however, provides protection against extremes of temperature, desiccation and ultra-violet radiation.

Thanks to the recent publications of Mani (1962; 1968), a considerable amount of information is available on the ecology and biogeography of high altitude arthropods. These are highly specialised and present a number of characters which, although by no means confined to high altitudes, provide a complex and unique combination associated with an equally unique combination of environmental conditions not met with elsewhere.

Melanism. Pronounced pigmentation of the body is one of the most striking characters of nearly all high altitude insects. In addition to black and brown, other colours commonly met with are red, orange and deeper tones of yellow than usually observed at lower levels. With the exception of cryptozoic forms, pale or colourless species are absent above the snow line. Moreover, the same species tends to be darker above the snow line than within the forest zone, or at lower elevations. As in Arctic snowlands (p. 14) its function is probably thermal.

Aptery. Reduction of wings or a totally wingless condition is characteristic of high altitude insects in all parts of the world. Mani (1968) lists many examples. In the mountains of East Africa (Salt, 1954) and Ethiopia (Scott, 1952; 1958), for example, many of the insects are flightless (Fig. 23) and restricted to particular massifs. Carabidae and Staphylinidae predominate. Wing atrophy can be associated with very diverse conditions including sex dimorphism, sedentary modes of life, myrmecophilous or termitophilous habits, parasitism and so on. In the case of the insects of high mountains, the explanation of aptery has been based, by analogy with the similar loss

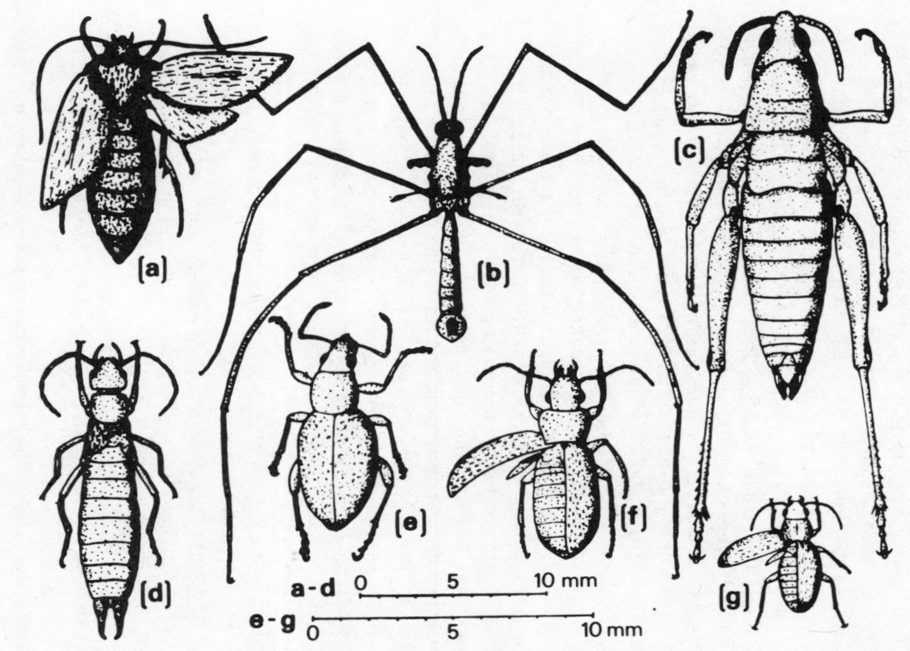

Fig. 23 Flightless insects found on Kilimanjaro above 4,250 m (14,000 ft): (a) *Saltia*; (b) *Tipula*; (c) *Parasphena*; (d) *Forficula*; (e) *Parasystiella*; (f) *Plocamotrechus*; (g) *Peryphus*. (From Cloudsley-Thompson, 1969 *after* Salt, 1954).

of flight in beetles of oceanic islands, on the supposed windiness of high altitudes. It might, however, equally be sought in the cryptozoic habit of the animals when they shelter from other violent conditions on the surface of the ground – from excessive radiation, great saturation deficiency, and wide and rapid fluctuations of temperature (Salt, 1954).

Size. Insects that live at high altitudes tend to be smaller than those from lower levels. Among Coleoptera, for example, relatively large species of *Calosoma* are not usually found at elevations much above the tree-line, even on the Himalayas where nearly every type of insect occurs at higher elevations than on most other mountains. Smaller species, however, have been collected on the highest elevations at which beetles have been found. The inverse relationship between body size and elevation has been described among Carabidae by Thiele and Kirchner (1958) and many other examples are cited by Mani (1968).

Not only is there a progressive reduction in mean body size with increase in elevation in most orders of high altitude arthropods, but orders that, like Diptera and Collembola which include very small forms, gain increasing dominance at higher altitudes and replace almost all others in the nival zone above the permanent snow line.

Insulation. According to Reinig (1932), not only is there a significant reduction in the mean body size of Tenebrionidae with increase in elevation but, at the same time, there is a progressive general flattening and increase in the width of the body, associated with an increase in the convexity of the elytra. The arching and flattening of the elytra in species from the high Pamirs is apparently to be correlated with differences in the prevailing ambient temperature. As in desert insects (p. 194), the enlarged sub-elytral cavity presumably contains a reservoir or air which insulates the body of the insect from excessive heating through insolation in the rarified atmosphere.

As elevation increases, there is an increasing preponderance of species with dense bristles, setae, scales and waxy coatings, especially among insects that frequent open snow and rock surfaces during the hours of bright sunshine. A dense clothing of bristles is found in Anthomyiidae (Diptera) of the Alps (Lindner, 1956) and of the north-western Himalayas (Mani, 1962). It probably serves not only to insulate the body from excessive heat and cold, but may also assist in the absorption of solar radiation.

According to Lindner (1956) the Anthomyiidae and other muscoid flies of the nival zone of the Alps show a characteristic forward bulging of the face or frons which is believed to increase the effective heat-absorbing surface of the body and thus also to aid in searching for warm patches of rock.

Cold stenothermy. Most high altitude arthropods are characterised by their pronounced stenothermy. They have a relatively low range of optimal temperature, usually around freezing point, and all their metabolic activities are adjusted to low temperatures. Most species, such as the larvae of Ephemeroptera, Plecoptera, aquatic Diptera and Collembola, which are adjusted to fairly uniform niche temperatures, develop normally under temperatures ranging from -1.5° to 5°C (29°-41°F) and are killed by a few minutes' exposure even to the warmth of a human hand (Mani, 1962). The cold stenothermy of high altitude insects varies, not only from species to species, but within succeeding stages in the life cycle of the same species. An insect that hatches from an over-wintering egg or re-awakens from hibernation, develops at increasingly higher temperatures with the advance of summer. There is, therefore, a gradual fall in cold stenothermy as the summer advances and the life cycle progresses from the dormant to the active stage (Mani, 1968).

In addition, the Arthropoda of the nival zone are heavily concentrated near the edge of the winter snow line, the ice margin or glacier snout, and the edges of glacial lakes and ponds. Only in moist situations can they expose themselves in the open and absorb warmth from the direct sunshine without risk of desiccation.

At high altitudes, life is severely restricted to the summer months. Seasonal rhythms of activity are therefore extremely well marked, as in the tundra and Arctic, except on equatorial mountains. Here there is no pronounced seasonal rhythm but all activity ceases at dusk and is resumed again at sunrise. In many species, especially of Diptera, maximum activity occurs between 08.00 and 11.00 hours. The period of activity of Lepidoptera is longer and, in the Himalayas, extends from 09.00 to 16.00 hours (Swan, 1961).

Prolonged hibernation under snow-cover, and a correspondingly short period of rapid development during the brief summer, are characteristic of high altitude insects of temperate and Arctic regions. Most species have short life-cycles, which are completed within one year, so that hibernation always takes place at the same stage of development. A few require longer than one year to complete a single generation at high altitudes: in general, however, the environment favours univoltine species wich pass through one generation per year. Insects with facultative diapause may complete two generations per year, while those with obligatory diapause are strictly univoltine (Lees, 1955).

It is curious that the insects which survive in the most inhospitable places should be members of the oldest and most primitive orders. Perhaps there is an analogy between the highest altitudes and latitudes and the sterile land of the Silurian period when terrestrial arthropods evolved from their aquatic ancestors. The first land animals may have fed on wind-blown debris that accumulated among the barren rocks of the world beyond the fringe of shore plants, just as high altitude Collembola, Thysanura and so on, subsist on pollen and plant debris blown up from lower levels (Cloudsley-Thompson, 1965d).

Adaptations of montane vertebrates

Montane animals show many adaptations to their environment, and the faunas of the highest peaks are very similar. The beetles of Kilimanjaro belong to the same genera as those of the Alps (Alluaud, 1908): and the high mountains of Africa, although widely separated, have similar bird faunas (Moreau, 1966). The high mountains of Java and the Phillipines present striking relationships with the Alpine animals of Asia. Speciation has, however, often resulted from isolation, as Hesse, Allee and Schmidt (1951) point out. The ibexes of the Old World, for example, fall into a series of species through the Alps, Caucasus, Ethiopian and Central Asian mountain ranges. Chamois *(Rupicapra rupicapra)* and wild sheep develop distinct sub-species in individual ranges. The same is true of many birds such as the partridges *(Caccabis* spp.*)* of Eurasia, the hedge sparrows *(Prunella* spp.*)* and snow finches *(Montifringilla* spp.*)*

Mountain dwelling mammals need to be sure-footed. The Nubian ibex *(Capra nubiana),* for instance, scales almost sheer faces in the mountains of Arabia and North Africa. When climbing, ibexes grip with the toes at the back of the feet. These are known as 'dew claws' and are not used on level ground. In hoofed mammals that do not inhabit mountain tops they are vestigial. Homeothermic inhabitants of high mountains tend to be larger than their relatives of lower levels but this is a response to cold (p. 193) rather than to altitude.

High altitude melanism is not confined to insects, but is notable also among mountain vertebrates. In the Alps this phenomenon is illustrated by *Salamandra atra, Lacerta vivipara, Vipera berus,* various rodents and other mammals. Although correlated with increased humidity, as Hesse, Allee and Schmidt (1951) point out, dark coloration is probably concerned functionally with the absorption of radiant heat.

Low atmospheric pressure. Vertebrate animals vary in their resistance to reduced atmospheric pressure. Their physiological adjustments have been surveyed by Folk (1966) who also makes a comparison between Arctic and Alpine environments. Perhaps the most important consequence of high altitude is low atmospheric pressure. The rate of pulmonary ventilation increases in response to low oxygen pressure and it is the circulatory system which is a limiting factor for newcomers at high altitudes. Animals living on high mountains have a high concentration of myoglobin in their muscles and their blood contains more haemoglobin per unit volume than usual.

A slight decrease in oxygen saturation of the arterial blood depresses the efficiency of the rods in the retina so that more light is required for normal vision. There is an alteration in the balance of acid to base in the blood, due to increased ventilation of the lungs, and the oxygen debt of men working to exhaustion decreases with altitude. These and other physiological consequences of hypoxia, or low oxygen tension, have engendered consumerate adaptation in the mammalian and bird faunas of high mountains.

The occurrence of reptiles and amphibians at high altitudes is little known and their physiology at low atmospheric pressure has been studied even less. Hock (1964b) concluded that hypoxia is not a factor limiting the occurrence of reptiles and amphibia at high altitudes. The availability of food and water, however, is a serious limiting factor for some species, while the daily and seasonal occurrence of temperatures adequate for food gathering, reproduction, and other activities are also important.

11 MICROENVIRONMENTS

Most of the animals inhabiting the earth are 'cryptozoic', and lead hidden lives in microenvironments where the conditions are favourable for their survival. They comprise an assemblage of small terrestrial forms, found dwelling in darkness beneath stones, rotten logs, the bark of trees, and other similar situations (Savory, 1971). The cryptosphere in which they live, includes fallen leaves that lie decaying on the earth's surface, twigs, stones, pieces of bark and fragments of rock and wood.

Another microenvironment, inhabited by numerous worms, arthropods and so on, is the soil (Wallwork, 1973). A distinction can be drawn between cryptozoa and subterranean, soil animals which represent different ecological groupings. Various other micro-environments support their own particular faunas. For instance, one of the most remarkable of micro-habitats in tropical forest is provided by 'reservoir plants'. These include the banana and *Heliconia* spp. which carry a little water in the axils of their leaves or bracts. The most interesting reservoirs, however, are those of the Bromeliaceae. Many of these are epiphytes, living on forest trees at all heights above the ground. The leathery leaves are arranged in a stiff rosette around a cavity which, even in the dryest weather, contains a quantity of water. This contains water plants such as mosses and bladderworts, green algae and diatoms. Organic matter is provided by dead leaves and drowned insects. Animals range from Protozoa, rotifers and minute crustaceans, to worms, water beetles, and even crabs. But the most abundant animals, usually, are the larvae and pupae of mosquitoes and chironomid flies, dragonfly nymphs and other insects. Several kinds of frogs are also found in the water and some of them breed there (Richards, 1970).

In a clump of bromelias in Trinidad, Scott (1912) found two frogs, a millipede, some woodlice, dragonfly larvae, cockroaches, an earwig, three species of beetle, a thrip and several Hemiptera. In Costa Rica, Calvert and Calvert (1917) obtained the following forms from a single bromelia plant: a scorpion, a false-scorpion, two species of Opiliones, eleven species of beetles and their larvae, a dragonfly larva, a stratiomyid larva, two species of Hemiptera, a caterpillar, an earwig, numerous ants and an earthworm.

Other important microenvironments include flowers and seed-pods in which many species of insects, mites and spiders commonly rest. Leaf-rollers and other insect larvae fasten leaves together to make nests. Ants of the genus *Oecophylla* co-operate in holding quite large leaves together while they are being bound in place with the silk produced by the larvae which are used as living shuttles. Another interesting plant micro-habitat is found in the ant-plants described in Chapter 3. Birds' nests, mammal burrows, holes in the ground, cliffs and walls are other important microenvironments for small animals (Cloudsley-Thompson, 1967a).

Microclimates

From a meteorological viewpoint, microclimates are considered in respect of small areas, a few kilometres in extent, or as the climate near the ground below standard weather-instrument shelters (Caborn, 1973; Geiger, 1950). This category of microclimate has sometimes been called 'ecoclimate' (Uvarov, 1931). From a zoological point of view, however, microclimates are usually regarded as something on an even smaller scale, such as the conditions under the bark of a tree, in ants' nests, or in burrows in the soil (Cloudsley-Thompson, 1967a; Smith, 1954). These categories of microclimate tend to grade into one another, but it is the latter with which the present chapter is concerned. (See p. 22)

Organisms occupying the same general habitat may actually be living under very different physical conditions. Many small terrestrial invertebrates, such as planarians, earthworms, woodlice, centipedes, millipedes, Collembola and other cryptozoa, avoid desiccation by remaining most, if not all of the time in a damp or humid environment (Cloudsley-Thompson, 1960a; 1968). Others, such as insects, spiders and mites, possess a thin epicuticular layer of wax which is relatively impervious to water vapour, and thereby reduces water loss by transpiration to a minimum.

Desert microclimates

The significance of microclimates is particularly marked in the case of desert animals, many of which inhabit holes or cracks in rock where they are insulated from the sun's heat. Violent diurnal temperature fluctuations are characteristic of the upper layers of the soil in deserts. A diurnal range of 56.5°C (101.7°F) has been recorded at a depth of 0.4 cm in Arizona. The diurnal range is so nearly the same as the yearly range, however, that the effects are not transmitted much below the surface and a relatively moderate and constant temperature is reached at a depth of 100–200 cm, which is well within the range of burrowing animals (Cloudsley-Thompson, 1964a; Sinclair, 1922).

High soil-surface temperatures are extremely superficial and do not penetrate for more than a few centimetres. Below 50 cm there is hardly any diurnal temperature variation in the sands of the Sahara and, at a depth of 1 m, the annual variation is not more than 10 deg.C (18 deg.F)

(Pierre, 1958). A diurnal temperature range of 30.5°C (55°F) was recorded on the surface of the sand in southern Tunisia in April. Five cm down a cricket's hole, the range was reduced to 18 deg. C (32.5 deg.F) while 30 cm down the hole, the temperature range was only 12.5 deg.C (22.5 deg.F) (Cloudsley-Thompson, 1956a).

Observations on desert microclimates in the Red Sea hills and coastal plain (Cloudsley-Thompson, 1962c) show that, given a very small reduction in conditions of extreme heat or dryness, an ecological chain can hang on even to such superficially unrewarding matter as dried vegetable matter (Brinck, 1956). Animal distributions tend to be influenced by extremes rather than by means and, in one series of *jebels* (hills), a surface temperature of 83.5°C (182.5°F) was recorded. The only animal to be seen was a solitary grasshopper. Four hours later, when the sand temperature had dropped to about 40°C (104°F), some ants ventured out. Low humidities were recorded, even among grassroots, and the primary biological advantage of vegetation apparently lay in the reduction of temperature it afforded. Elsewhere, invertebrates survived only by burrowing deeply (Cloudsley-Thompson, 1962c).

An outstanding feature of the Saharan sands is their relatively high humidity. Even during summer, the air surrounding loose grains at a depth of only 50 cm has a relative humidity of 50 per cent, according to Pierre (1958). This moisture arises from the water, presumably derived from rainfall in the Atlas, Hoggar and Tibesti mountains, and from beyond the desert limits, which underlies much of the Sahara. Moreover, water descends very quickly through wind-blown sand because its permeability is high and anti-wetting prioperties low. Owing to capillary tension, a given charge of water applied at the surface of dry sand will sink to a certain depth and no more. This depth is about eight times greater than the immediate precipitation. Water that has reached a depth of 20-30 cm remains as a moist, unsaturated zone for several years because, sand being a very poor conductor of heat, the temperature is relatively constant and there is no ventilation. The sand above and below is dry (Bagnold, 1954).

Not only is protection from climatic extremes obtained beneath rocks and in deeper layers of the soil, but caves and rock clefts form a natural habitat for many desert animals (Buxton, 1923; Cloudsley-Thompson, 1956; Williams, 1954). The presence of plants also provides considerable protection from climatic extremes in hot-dry environments. Consequently, within distances of a relatively few metres there may be available, in arid country, a wide range of microclimatic conditions from which an animal can select the most suitable for its requirements with little expenditure of energy. By burrowing deeply or entering a cave, the extreme heat of the daytime can be avoided; and, by leaving the burrow at night, the peak temperature below ground can also be avoided, for there is a considerable time-lag before heat begins to penetrate deeply into the sand (Cloudsley-Thompson, 1956a).

Arctic microclimates

The microclimates inhabited by Arctic plants and animals have been discussed by Corbet (1972), who concludes his important review by pointing out that the biologically significant features of Arctic microclimates, as of Arctic climates, relate to the seasonal and diel, or daytime patterns of insolation. Towards the pole, insolation becomes progressively restricted to the short summer. The contrast between winter and summer therefore becomes greater.

In favourable microhabitats, there is an amelioration of the microclimate near the ground which warms abruptly under direct insolation. In sunshine, plants and animals inhabiting this zone raise their own temperatures yet further by their structure, posture and orientation. These two circumstances may give plants and, even more, animals, access to temperatures at least 20°C above current air temperatures. At 83° North, on a clear, calm summer's day, the body temperature of an insect forcibly confined to the microclimate on the surface of a plant might well exceed its upper lethal limit.

Wind is the main factor that cools insolated objects. Arctic organisms are unusually exposed, not because the wind is particularly strong, but because there are few obstructions to check its flow. The layer of air nearest to the ground is warmest, partly because the cooling effect of the wind is lowest at ground level. The factors providing an optimum microclimate in Arctic regions are: exposure to sunshine, shelter from wind, the presence of hills (especially to the north) to absorb counter-radiation, and the absence of shade. A south-facing aspect, ground of low thermal conductivity and the presence of ponds and lakes without marginal shade and with gently sloping shores are also beneficial.

Two desirable features for winter survival are early, heavy snowfall that is well distributed over level and south-facing ground and which melts quickly in spring, and the presence of moderate-sized lakes not more than 2 m deep (Corbet, 1972).

The principles involved in this discussion of desert and Arctic microclimates apply equally well, although to a less marked degree in other, less extreme, biomes. Undoubtedly one of the best general accounts of micro-climates, and the measurement and recording of their physical components, is that of Macfadyen (1963).

Fauna

The evaporating power of the air is the most important physical factor of the environment affecting the distribution of cryptozoic animals.

This is because small creatures have a very large surface area in proportion to their mass. The conservation of water is, therefore, the major physiological problem of their existence. This is achieved mainly through behavioural responses that restrain them to humid environments, for most cryptozoic forms appear to lack a waterproof integument (Cloudsley-Thompson, 1967a).

Cryptozoic and soil-dwelling animals include members of most, if not all, terrestrial phyla, viz. Protozoa, Turbellaria, Nematoda, Gastrotricha, Nemertini, Annelida, gastropod molluscs, and Arthropoda (Fig. 24). The biology of soil animals has been discussed in many recent books, of which the following merit special mention: Kevan (1962), Kuhnelt (1961), Russell (1957), Shaller (1968) and Wallwork (1970). Cryptozoa have been considered by Cloudsley-Thompson (1967a), Lawrence (1953) and Savory (1971), while special aspects of the subject are treated by Allee *et al.* (1949), Cloudsley-Thompson (1969), Hesse, Allee and Schmidt (1951), and Macfadyen (1963).

Apart from their small size, which conveys both advantages and disadvantages, cryptozoa tend to have thin exoskeletons. Many of them are blind, and the sense of vision is largely replaced by tactile and chemotactic senses and a low degree of elaboration. These characters have been discussed by Savory (1971).

Cryptozoic animals are usually completely dependent upon a moist environment. Many, such as planarians, nematodes and earthworms, have moist body surfaces and therefore naturally lose water rapidly through transpiration. Among the Arthropoda, woodlice, centipedes, and millipedes belong to classes whose orders do not normally possess epicuticular wax layers. In them, water loss by evaporation therefore tends to be proportional to the saturation deficiency of the atmosphere (Cloudsley-Thompson, 1960b; 1961a). Most insects and arachnids possess comparatively impervious integuments, as we have seen. Nevertheless, cryptozoic and soil-dwelling representatives of these classes tend to lose water quite rapidly in dry air (Cloudsley-Thompson, 1969).

Powers of locomotion are not well developed among the cryptozoa, for they seldom need to move far from one part to another of their uniform environment. Wingless insects are common, and those that do possess wings are not strong fliers. Relatively inactive forms, living for the most part in darkness, neither require nor possess well-developed eyes. The sense organs most used are tactile hairs on the antennae and limbs. While most cryptozoa feed on decaying vegetable matter, fungi, and so on, a few are carnivorous and find their prey chiefly by the sense of touch. Examples are provided by false-scorpions (Chelonethi) whose large claws bear many long tactile hairs, ground-spiders of the families Gnaphosidae and Clubionidae, in which the sensory hairs are

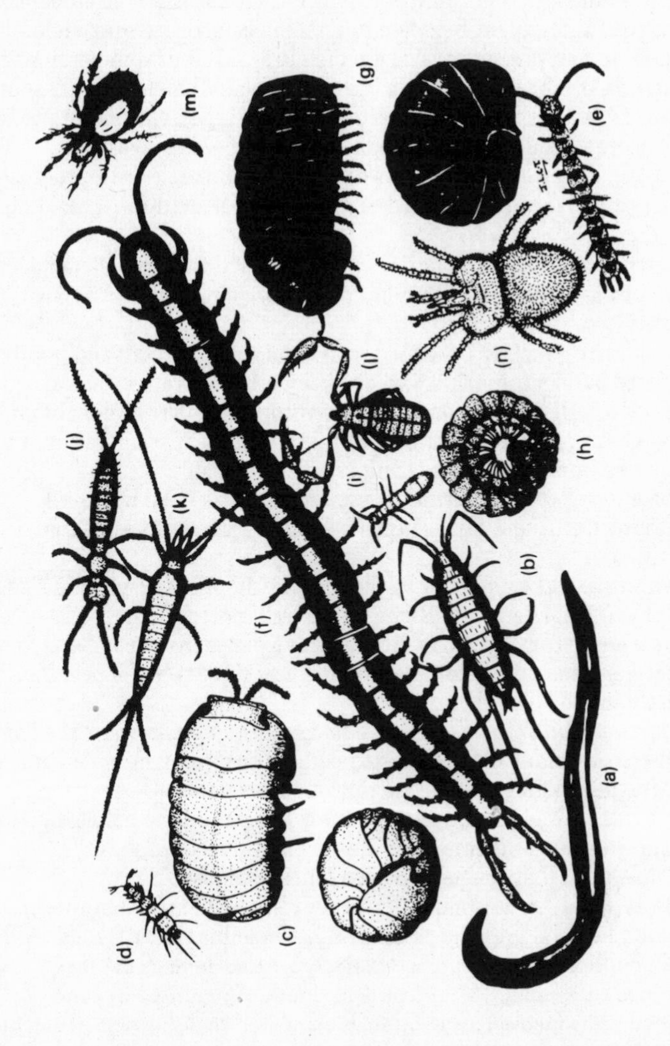

Fig. 24 Cryptozoic animals: (a) Land planarian *(Artiocotylus)*; Amphipod *(Talitroides)*; Pill-woodlouse *(Cubaris)*; (d) Pauropod; (e) Symphylid; (f) Centipede *(Scolopendra)*; (g) Pill-millipede *(Sphaerotherium)*; (h) Colobognath millipede *(Cylichnogaster)*; (i) Collembolan; (j) Dipluran *(Campodea)*; (k) Thysanuran *(Machiloides)*; (l) False-scorpion *(Cheiridium)*; (m) Oribatid mite; (n) Trombiculid mite. (Not to scale.) (From Cloudsley-Thompson, 1969.)

carried on the legs, and centipedes which carry them on the antennae.

Like parasites, plant-bugs and other animals that live surrounded by food, the cryptozoa are sluggish creatures, without well-developed respiratory mechanisms. They cannot withstand drought, yet are soon drowned if their environment becomes water-logged. Such contingencies are rare, however, in the cryptosphere, because a short, vertical migration, either upwards or downwards, can soon remove a cryptozoic animal from danger. Specialised modes of life are neither required nor have they been evolved.

Lawrence (1953) summarises as follows the distinctive characters of cryptozoa as compared with other groups of animals of the African forests.

(a) *Small size* which is associated with absence of pigment and poorly chitinised cuticle through which considerable evaporation can take place.

(b) *Lack of efficient respiratory mechanisms* which are either absent or, when present, lack devices that control water-loss by evaporation. For this reason, most cryptozoic animals are nocturnal.

(c) *Poor development of visual sense organs,* associated with well developed tactile and taste sensillae. Auditory organs have seldom been identified.

(d) *Primitive and archaic forms* such as Onychophora, Pauropoda, Symphyla, Palpigradi, and Ricinulei are well represented.

(e) *Secondary sex characters* distinguishing males from females are much reduced and there is usually little in the way of outward appearance to separate the sexes.

(f) *Development from eggs that do not possess an impervious shell* and are therefore laid in clusters coated with mucus. In many orders the delicate, thin-skinned young are retained for a while after hatching in a brood-chamber or its equivalent. During this period the helpless young remain with the mother and are covered with the secretions she produces until they are able to exist independently.

(g) *Moulting* often continues throughout life, even after maturity has been reached and growth ceases. Striking examples are afforded by the Onychophora and Diplopoda. Many cryptozoic animals eat their discarded integument, a type of behaviour not practised by land-adapted arthropods. This habit may be related to a shortage of calcium in tropical soils.

(h) *Activity.* Most cryptozoa are inactive, slow-moving animals that spend long periods of time in a passive state. Scutigeromorph centipedes and the Palpatores among the harvest-spiders are exceptional in being speedy but, even among these, sustained displays of activity are infrequent.

(i) *Social life* does not exist, although some species are mildly gregarious (Lawrence does not regard termites and ants as belonging to the

cryptozoa) and parasitism is not a feature.
(j) *Lack of diversity of form and colour.* Genera, families and even orders consist of a small number of species which display considerable uniformity of appearance and structure. The Collembola and Acarina are probably the richest in diversity of all the cryptozoa, yet the number of species contained in their respective orders falls far short of that of many families of pterygote insects.
(k) *The family history of the cryptozoa* has greater continuity and reaches back longer in time than does that of the present inhabitants of open lands. Many cryptozoic orders of insects were already established in the Palaeozoic period and only a minority has failed to survive until the present. All this suggests a community of animals which, in former times, was the product of an environment more universally simple and uniform than it is today. Evolutionary progress has died away leaving the cryptozoa imprisoned in an environment from which no advance can be made and from which escape is forever impossible.

Diurnal rhythms

Just as plants and their habitats are closely related, so are small animals and the microenvironments they inhabit. Although it is possible for animals to survive for a while in unfavourable circumstances, they differ from plants in their ability to move and seek out more favourable resting sites. For this reason, the influence of microclimatic conditions and their influence on diurnal and seasonal rhythms of activity are of extreme importance in the ecology of small animals.

Cryptozoic and soil-dwelling animals are essentially nocturnal in habit. Restricted during the day to sheltered, moist micro-habitats, they emerge at night to crawl over the surfaces of stones, bare ground, tree trunks and so on. In the case of the woodlouse *(Oniscus asellus)* a circadian pattern of locomotory activity has been shown to result from a combination of an inherent rhythm and a direct response to environmental changes. Activity is correlated with alternating light and darkness, rather than with fluctuating temperature and humidity. The photonegative response of these animals increases during darkness, ensuring a prompt return to sheltered conditions at dawn. On the other hand, woodlice become photopositive in dry air so, if their daytime microhabitat should dry up, they are not restrained there by the daylight but are able to wander in the open until they find some other damp hiding place when they again become photonegative (Cloudsley-Thompson, 1952). The intensity of nocturnalism is related to the rate of transpiration in woodlice (Cloudsley-Thompson, 1956b).

From his investigations on populations of *Armadillidium vulgare,* Paris (1963) likewise concluded that greater activity occurred above ground when the saturation deficiency of the atmosphere was low. In apparent contradiction to these findings, Den Boer (1961) suggested that surface activity of *Porcellio scaber* was greatest when saturation deficit was high, although vertical movement was negatively correlated with saturation deficiency. In attempting a synthesis of these conflicting observations, Wallwork (1970) suggested that surface activity may subserve different functions in different species. Secondly, the response of an animal to a given situation may vary with its physiological state. Edney (1968) pointed out that the surface activity of the population of *A. vulgare* observed by Paris (1963) was undoubtedly associated with feeding, and that the animals involved were probably not overhydrated; whereas the activity of the *P. scaber* described by Den Boer (1961) was a behavioural device to remove from the body excess water absorbed during the day when the woodlice were in a moist, subterranean micro-habitat.

The relationship between circadian rhythm and microenvironment is better understood in woodlice than in other cryptozoic animals. Nevertheless, it is quite probable that similar relationships may exist in other arthropodan groups of the soil and cryptosphere.

Seasonal rhythms

Reproductive cycles have been determined in several species of woodlice: the literature is reviewed by Cloudsley-Thompson (1975a). An endogenous rhythm in *Porcellio scaber,* associated with the production of hormones, has been shown to control reproduction. The rhythm is associated with photoperiod (Wieser, 1963). A seasonal rhythm in the release of ammonia is also apparent in this species and in *Oniscus asellus* according to Wieser, Schweizer and Hartenstein (1969). As in the case of many tropical animals, on the other hand, the breeding of *Periscyphis jannonei* in the Sudan appears to be controlled not by photoperiod but by the rains (Cloudsley-Thompson, 1975a).

12 FRESH WATERS

Fresh water rivers, lakes and inland seas comprise innumerable bodies of water, varying in size and depth, and spread across the continents of the world. Most of them are comparatively isolated, and they persist for relatively short periods of geological time (Beaufort, 1951).

Compared with the sea, even the deepest of these inland waters are shallow. Only Lake Baikal (1,706 m) and Lake Tanganyika (1,435 m) have a depth of over 1,000 m, whereas the average figure for the oceans is about 3,795 m, according to Hesse, Allee and Schmidt (1951). Indeed, few lakes are deeper than 400 m and most of them are under 100 m. Lakes, ponds and pools grade into marshes, and a similar gradation is found in running waters. Consequently, the extent of shore line and shallow bottom is relatively very much greater in inland waters than it is in the sea. The ratios of the extents of habitats suitable for sessile and free-swimming forms are therefore very different. Reid (1961) divides inland waters into lakes, streams and estuaries. Only the first two of these categories are considered in the present chapter. The most important books on limnology are those of Hutchinson (1957; 1967); lesser texts include Carpenter (1928), Macan and Worthington (1951), Reid (1961), Ruttner (1953), and Welch (1935). A comprehensive work on the flora and fauna of plankton is by Ward and Whipple (1959). See also Allee *et al.* (1949).

Hydroclimate

The major terrestrial biomes are dependent upon climate – a complex of physical factors. By analogy, the combinations of physical and chemical factors in aquatic environments are often referred to as hydroclimates (Wasmund, 1934).

The chief hydroclimatic factors are as follows:

(a) **Density.** The density of water is nearly one thousand times greater than that of air, and it increases with depth and salinity. The density of protoplasm is slightly greater than that of sea water so that there is a tendency for living organisms to sink to the bottom unless equipped with adaptations for floating. This tendency is, of course, greater in fresh than in saline waters. Strengthening of the body for support and protection, so necessary on land, is not essential in water. This makes possible a greater variety of structure in aquatic than in terrestrial environments (Hesse, Allee and Schmidt, 1951).

(b) **Pressure.** The pressure of water at great depths is enormous, but it is doubtful if this factor is responsible to an appreciable extent for the paucity of life in deep water because all the tissues are permeated by the medium. The absence of light and food is of much greater biological significance.

(c) **Salinity.** The salt content of the sea is relatively uniform, but that of fresh waters may vary by a factor of more than 20. Salinity affects plants and animals, not so much through its influence on the density of the water, as by the osmotic pressure it engenders. The osmotic pressure of sea water is over 20 atmospheres, but that of fresh water is relatively low (Clements and Shelford, 1939). The amount of sodium chloride varies among inland waters from 14 ppm in the Rhine to 2,313 per cent in the Dead Sea. In some waters, such as the salt seas of the steppes, the content changes with the seasons. The amounts of magnesium and calcium carbonate also vary; so that water may be 'hard' or 'soft', depending upon the content of lime. Fresh-water sponges, Bryozoa and certain Crustacea are intolerant of excess calcium, whose presence is essential to many snails and lamellibranch molluscs. Other inorganic substances may also be present in great quantities and are tolerated in varying degrees by the flora and fauna. These include poisonous compounds such as hydrogen sulphide, sulphur dioxide, borax, and salts of iron.

(d) **Acidity.** The pH of fresh water may have a considerable influence on its flora and fauna. The waters of bog lakes are usually dark coloured, acid and sparsely populated. In contrast, the high concentration of carbonates and a low content of carbon dioxide, characteristic of many hill streams, constitutes a more favourable environment for plants and animals. Reid (1961) summarises the relationships between carbon dioxide, bicarbonates and carbonates in relation to pH in lakes and streams as follows:

- (i) The pH value varies inversely as the concentration of dissolved carbon dioxide, and directly as the bicarbonate concentration.
- (ii) The critical value relating to the presence or absence of free carbon dioxide is pH8, the free gas being absent above that value.
- (iii) The absence of free carbon dioxide does not limit photosynthesis of certain algae and higher plants, some of which are adapted for utilisation of carbon dioxide from carbonates.

(e) **Temperature.** Like other environmental factors, temperature tends to be more uniform in aquatic than in terrestrial environments. This is because the specific heat of water is about 500 times greater than that of air, and the thermal conductivity of water is approximately 30 times greater. Aquatic vegetation types are, therefore, more widespread than terrestrial ones, although the ecological differences between terrestrial and aquatic habitats are largely a matter of degree. Chemical and physical conditions in the soil are extremely important to terrestrial plants, and aquatic vegetation is influenced more conspicuously by light than by temperature.

The most outstanding effects of temperature on lakes are concerned with thermal stratification. As the temperature approaches 4°C, from above or below, the density of water increases: with further cooling or warming it decreases. Consequently, in winter when the surface is frozen, lakes are stratified inversely with colder, lighter water nearer the surface, resting on the denser water which has a uniform temperature about 4°C. In spring, when the surface waters become warmed to 4°C, they have the same density as deeper water so that the wind is able to cause mixing. As the temperature rises, the surface water is warmed above the point of maximum density, rests on the colder denser water below, and the lake is again stratified. On the degree and permanence of stratification depends the oxygen content of the deeper waters of lakes.

The *epilimnion* or surface waters of lakes have variable temperatures. Stirred by the wind, with plenty of light and oxygen, they are rich in plankton. The epilimnion extends to a depth of 10 m or so. Below it is a transitional area or *thermocline* in which the temperature shows a maximum change with depth. The deepest waters or *hypolimnion* experience a maximum annual temperature range seldom greater than 5°C. There is no light, and oxygen may be deficient, so plants and animals are scarce (Fig. 25).

According to their thermal properties, lakes can be classified into the following types:

(i) *Meromictic.* These are lakes in which there is no mixing of deep and shallow water at any time. They occur in the tropics, where the temperature varies little from one period of the year to another, and are deep. Lakes Tanganyika, Malawi and Kivu, for example, are meromictic and more or less permanently stratified. The lower layers are anaerobic so that methane and hydrogen sulphide are produced.

(ii) *Oligomictic.* A very small temperature difference between the surface and bottom suffices to maintain stable stratification in these lakes so that the water circulates only at rare intervals when abnormal cold spells occur. Oligomictic lakes are likewise found in tropical regions and may be either small or moderate in extent, or of very great depth, or in regions of high humidity.

(iii) *Monomictic.* These circulate once per annum. Warm monomictic lakes occur in the tropics and circulate in winter at temperatures above 4°C. Cold monomictic lakes, on the other hand,

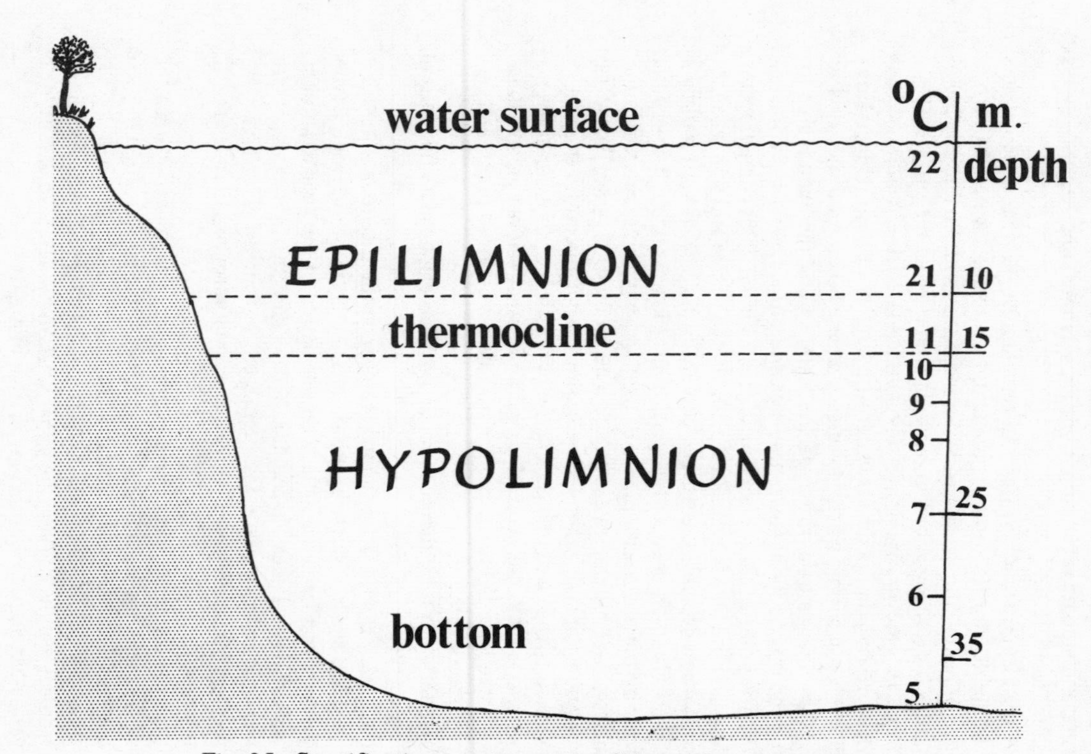

Fig. 25 Stratification in a temperate climate lake in summer.

circulate in summer at temperatures below 4°C.

(iv) *Dimictic* lakes circulate twice a year in spring and autumn. They are characteristic of temperate regions.

(v) *Polymictic.* These are usually large and shallow: they show no persistent thermal stratification and may be found in all parts of the world (Hutchinson, 1957).

(f) **Light.** Increasing depth of water is characterised by differences in plant life, which are related to decreasing light intensity. The temperature of the water, its chemical composition and aeration, may also be important locally, as already mentioned, but absence of light is more often a limiting factor for plant life. The three main zones, represented by different forms and abundance of life at decreasing levels of light intensity, are as follows:

(i) *Euphotic zone.* In this the light is relatively bright and large plants can develop normally.

(ii) *Dysphotic zone.* Here the light is dim and only small algae and mosses can exist.

(iii) *Aphotic zone.* Only non-photosynthesising organisms can exist at depths to which no light penetrates.

The limits of these zones are imprecise: they vary according to the angle of the sun, which affects the pentration of light, and to the turbidity of the water.

(g) **Oxygen.** The solubility of oxygen in water is affected both by temperature and by salinity. Even at its maximum possible solubility, however, oxygen is less than one-quarter as concentrated as in air. In nature, values probably range from a maximum of about 6 ml $0_2/1$ (3 per cent of the concentration in air), to nil in anaerobic conditions. Moreover, although fairly evenly distributed in the atmosphere, the concentration of dissolved oxygen varies considerably according to time and place.

The absence of oxygen is, surprisingly, not always a factor limiting animal life. In the lower levels of the meromictic crater lake Nkugute in western Uganda, for example, the water does not circulate and is devoid of oxygen. Neverthelss a copepod *(Thermocyclops schuurmanni),* rotifers *(Horaella brehmi* and *Keratella tropica),* as well as ciliate Protozoa, occur right down at the bottom, with some 40 m of oxygen-free water above them. It is probable that the eggs of *T. schuurmanni* are produced in oxygenated water and sink to the bottom where the first stages of development proceed in the absence of oxygen. Presumably the young copepods then return to the surface to mature and reproduce (Beadle, 1963).

The amount of oxygen in the deep water regions of lakes may vary considerably, depending upon the extent to which the upper and lower layers of water intermix, as we have seen. In very deep lakes there is a steady deposition of sediment at levels below those in which vegetation grows. Bacterial decomposition of this may result in reduction of oxygen near the bottom.

Flora

The vegetation of lakes can be divided into the phytoplankton consisting of minute floating plants such as diatoms and desmids, and benthos, the semi-aquatic marginal communities of vascular plants, attached or loose bottom-living Algae.
(a) *Phytoplankton.* The main categories of freshwater plankton are: (i) The *limnoplankton* of lakes and ponds, and (ii) the *potamoplankton* of slow streams and rivers. Depending upon the presence of dissolved minerals, bodies of fresh water may be: *(a) Oligotrophic.* Dissolved minerals are in low concentration, but desmids are abundant and there may be a narrow zone of rooted higher plants; *(b) Dystrophic.* Again the waters are poor in nutrients but rich in humus and acidic in reaction. They are often coloured, containing desmids and bog-mosses; *(c) Eutrophic.* These usually possess fewer species, but are richer in individuals, than are oligotrophic waters. They are relatively rich in compounds of nitrogen, phosphorus and often calcium and, typically, contain plentiful blue-green algae, a broad zone of rooted pondweeds, and a surrounding one of luxuriant reed-swamps (Polunin, 1960).
(b) *Benthos.* Units of vegetation, when left to themselves, tend to change from less complex communities of small plants to more complex ones dominated by larger plants of higher life-forms. The developmental stages of such successions are known as 'seres' and lead to a relatively stable climax. In temperate regions, the typical aquatic sere or 'hydrosere' begins in bodies of fresh water whose beds become colonised by aquatic vascular plants, mosses and algae as deeply as sufficient light penetrates. The plants form dense mats in which silt and humus accumulate until the bed is raised to a depth of 1–3 m, when it can be invaded by water lilies *(Nymphaea* spp.*)*, pond weeds *(Potamogeton* spp.*)* and other plants with floating leaves which suppress the submerged plants. Eventually the soil surface rises to a point at which shrubs and trees can become established, giving rise to hygrophytic woodland (Polunin, 1960).

Marshes, bogs and saline waters

Bogs and marshes, so common in the cooler regions of the northern hemisphere, form a special habitat in which the substratum is composed of waterlogged peat. Above this is a layer of mosses, especially *Sphagnum* spp., surrounded by heathland vegetation (Chapter 7). The vascular plants of wet depressions are chiefly hygrophytic sedges

(Carex spp.*)* and cotton-grasses *(Eriophorum* spp.*)* in both temperate and Arctic regions.

Bog waters tend to be strongly acidic, with a high content of humic materials which frequently impart a brownish coloration, and deficient in dissolved salts. They are characterised by a microflora rich in desmids, Cyanophyceae (e.g. *Chroococcus* spp.), diatoms, peridinians and other green algae.

Inland saline waters vary from faintly brackish to several times the average salinity of the sea, some being completely saturated with salts. Especially in desert regions, their flora consists mainly of green algae (e.g. *Dunaliella salina),* although diatoms and Cyanophyceae also occur. Under only slightly brackish conditions the flora is predominantly a freshwater one: in really saline waters it is mainly marine in type (Polunin, 1960).

Fauna

The flora and fauna of inland waters may be freely floating plankton, which includes microscopic plants (phytoplankton) and tiny animals (zooplankton), fishes and other larger swimming animals (nekton) and bottom-living forms (benthos). The zooplankton is composed mainly of Protozoa, rotifers and Crustacea, the benthos of Protozoa, Nemertea, Nematoda, Gastrotricha, annelid worms and insects.

Locomotion

The adaptations of planktonic animals for floating include a large surface to volume ratio, often enhanced by numerous spines that reduce the rate of sinking, gelatinous tissue, and floats which lower specific gravity. For example, *Leptodora kindti* (Fig. 26) is a predatory planktonic crustacean with numerous hairs and spines, large compound eyes and raptatory forelimbs. It swims with oar-like antennae. A gelatinous constituency is characteristic of pelagic jelly-fishes (Scyphozoa). Although these are mostly marine, medusae of *Craspedacusta sowerbii* (Fig. 26) are dispersed sporadically throughout the waters of North and South America (Reid, 1961). Related species are widely distributed throughout Europe and Asia, while species of *Limnocruda* occur in the lakes and rivers of Africa (Cloudsley-Thompson, 1969; Hutchinson, 1957).

Swimming can be achieved by throwing the body into waves, using the limbs as paddles, or by jet propulsion. The finest swimmers, such as fishes and river dolphins (Platanistidae), progress by twisting their bodies into a series of waves or undulations. The sideways movements of the tails of fishes are due to the alternate contraction and relaxation of the antagonistic muscles lying on either side of the backbone, while the

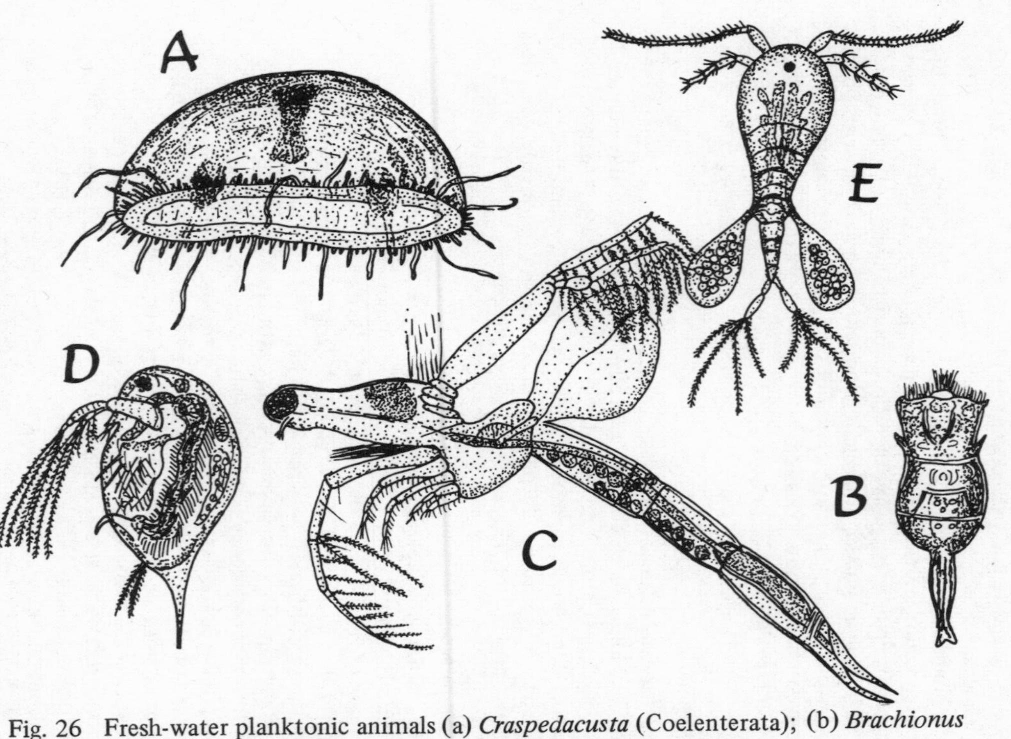

Fig. 26 Fresh-water planktonic animals (a) *Craspedacusta* (Coelenterata); (b) *Brachionus* (Rotifera); (c) *Leptodora;* (d) *Daphnia;* (e) *Cyclops* (Crustacea).

front part of the body provides a fulcrum about which the muscles turn the remainder of the body, including the tail and its fin, so that they push backwards and sideways against the water. As Gray (1968) and his co-workers have shown, an eel can glide over a board if this has been studded with nails or pegs so that the eel's body has something to press against. Pressure is exerted by the inner curves of the body, against water when the eel is swimming, as it is against the pegs. The movements of a speedy fish, such as a trout or salmon, are similar to those of an eel. Apparent differences are almost entirely due to differences in the relative lengths and flexibilities of the bodies.

Not only fishes have streamlined contours. The smooth, polished, elliptical shapes of many water beetles serve to reduce resistance during swimming; while the flattened form of mayfly nymphs and the larvae of certain Diptera and Coleoptera of torrential streams, associated with anchoring devices such as spines or suckers, helps to prevent such creatures from being swept away by the current.

Many aquatic insects swim with their limbs, which are used as paddles (whose surface area is often enlarged by a fringe of hairs). So, too, does the water spider *(Argyroneta aquatica)*, web-footed amphibians and turtles. Jet propulsion is found among larval dragonflies *(Odonata)*, and most Anisoptera larvae use it as a means of escape – weakly if they are burrowers, and powerfully if they are weed-dwellers. In larvae of *Aeshna* spp. the act of swimming comprises three processes: (i) expulsion of water from the rectum; (ii) rowing or beating movements of the legs against the abdomen, and (iii) pressing the legs against the abdomen as jet propulsion continues (Corbet, 1962).

There are several other methods of moving in water that have been adopted by insects, but most of them are modifications of those already mentioned. The minute aquatic fairy flies (Mymaridae) swim through the water by flapping their wings. Worm-like midge larvae (Ceratopogonidae) swim like eels by bending their bodies from side to side. Mosquito larvae do the same, but the chief propulsive force comes from a row of long hairs projecting downwards from the terminal segment of the abdomen (Wigglesworth, 1964), When beetles of the genus *Stenus* fall on water they expel from their anal glands a substance which lowers the surface tension, so that they are drawn rapidly forwards like a toy boat propelled by camphor. If the tip of the abdomen is removed, they can no longer move in this manner (Billard and Bruyant, 1905).

Very small animals are able to swim by ciliary action but, as soon as they reach a length of about 1 mm, the inertia of their bodies becomes too great for cilia to move them. Certain larger forms, such as planarians and some water snails, are able to glide by means of cilia while their mass is supported by the surrounding medium.

Respiration

Life in water is subject to conditions very different from those obtaining on land for, unlike terrestrial forms, aquatic organisms are in no danger of suffering from desiccation. On the other hand, as we have already seen, the concentration of oxygen in water is much lower than it is in air.

Many small, aquatic animals, including worms, Crustacea and insects, obtain oxygen by absorption through the body wall. Larger forms are usually equipped with gills. Among arthropods there are special areas of the body where a rich supply of tracheae is covered by thin, permeable cuticle so that oxygen can diffuse from the surrounding water into the tracheal system. Such areas may extend all over the surface of the body or they may take the form of leaf-like gills containing numerous tracheoles, as in the larvae of stone-flies (Plecoptera) and may-flies (Ephemeroptera). The gills of Zygoptera larvae are attached to the hind end of the abdomen while those of Anisoptera lie inside the rectum.

In many aquatic insects only the spiracles at the posterior end of the abdomen are functional and the terminal segment is drawn out to form a respiratory tube or siphon, with the spiracles opening at its end. This siphon is protruded into the atmosphere when its owner needs to breathe or, less frequently, into the inter-cellular spaces of aquatic plants such as the grass *Glyceria aquatica.* Among British insects, this remarkable adaptation is seen in the larvae of the mosquito *Taeniorhynchus richardii* and in the larvae of leaf-beetles of the genus *Donacia* (Imms, 1947). Insects that protrude their respiratory tubes through the surface of the water include the bugs *(Nepa* and *Ranatra* spp.*)*and larvae of aquatic Diptera (Culicidae, Tipulidae, Stratiomyidae) etc. An extreme example is afforded by the rat-tailed maggots of drone-flies *(Eristalis* spp.*)*. These have a highly extensible siphon which can be lengthened to 12 cm or more in a larva only 2 cm in length.

A fringe of semi-hydrofuge hairs around the spiracles enables contact to be made between the atmosphere and the air in the tracheae. Hydrofuge hairs are also employed to maintain a layer of air on the undersurface of the body of the silver water-beetle *(Hydrous piceus).* In *Dytiscus* spp., on the other hand, air is stored under the wing cases or elytra. Air stores may have an important hydrostatic function and are so arranged that the insect rises to the surface with the desired point uppermost, and the same is true of the air in the tracheal system of aquatic larvae (Frost, 1942; Imms, 1947; Miall, 1895; Wigglesworth, 1964).

A film of air on the surface of the body may not only have an hydrostatic function and serve as a store of air, but it may also serve as a gill which extracts dissolved oxygen from the water if it is in communication with the spiracles. As the oxygen in the air store is used up, an excess of nitrogen remains. In consequence, oxygen, which diffuses more rapidly than nitrogen, leaves the surrounding water and replaces that which has been consumed. Of course, some nitrogen is lost at the same time, but the process goes on as long as any of this gas remains.

Some insects have a permanent gill or plastron of this kind in which the air is held by a pile of fine hydrofuge hairs bent over at right angles at the tip so that they form a hydrophil surface covering a thin film of air (Thorpe, 1950). This arrangement is found in beetles of the genera *Haemonia* and *Elmis*, but the finest example is undoubtedly that of the aquatic bug *Aphelocheirus aestivalis* which can live permanently submerged in the Danube up to a depth of 7 m (23 ft.). The greater part of the body surface of this remarkable insect is covered with an extremely fine plastron held in position by an epicuticular hair pile having approximately 2,000,000 hairs per sq. mm (Fig. 27) (Thorpe and Crisp, 1947). It is extremely difficult to displace this plastron of air, even by pressures up to four or five atmospheres.

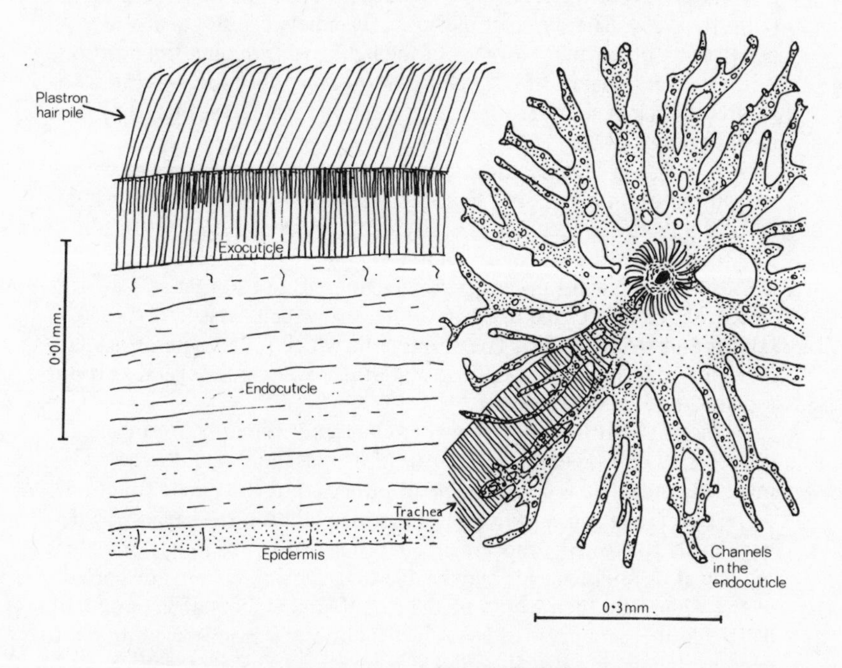

Fig. 27 Plastron of *Aphelocheirus aestivalis* (in section, highly magnified) and spiracular rosette. (After Thorpe, 1950.)

The respiratory current of fishes is produced by a sideways movement of the operculum which enlarges the bronchial cavity. The branchiostegal folds below the operculum prevent the inflow of water from behind and, when the operculum is moved inwards, the exit of water forwards is prevented by flaps in the throat. Many fishes are able to live outside the water and, especially among tropical forms, a variety of adaptations for breating air have been evolved. These are summarised by Norman (1963), Young (1950) and others.

Diurnal rhythms

Diurnal vertical movements are characteristic of planktonic organisms and occur among the fauna and flora of lakes and larger rivers. They are extremely well documented, especially in marine plankton but are also well known in inland waters (Carpenter, 1928; Cloudsley-Thompson, 1960b; Cushing, 1951; Reid, 1961). In Lake Windermere, for instance, the copepod *Cyclops strenuus* is in greatest abundance at some distance from the surface. As the day draws on there is an upward migration and, about midnight, the surface layers of the lake are most thickly populated. A return migration to the depths occurs at dawn (Fig. 28) (Ulyott, 1939). There is a strong correlation between the depth of penetration of the blue segment of the light spectrum and the position of greatest abundance of *C. strenuus.* Moreover, the copepods move to greater depths in summer when light intensity is greatest than they do in the spring.

A population of water-fleas *(Daphnia magna),* in a tank filled with a suspension of indian ink in tap water, will undergo a complete cycle of vertical migration when the intensity of the light is altered artificially. The function of the indian ink in this experiment is to reduce the amount of movement necessary for the animals to experience a significant change in the intensity of light to which they are exposed (Harris and Wolfe, 1955). The movements which follow alterations in light intensity are active responses, and the downward movement is not merely a passive sinking.

Although light intensity appears to be a controlling factor in the distribution of zooplankton, temperature gradations may also be involved, and there is considerable dispute as to the ultimate function of vertical migration. In the case of marine plankton the suggestion has been made that regular movements between horizontal layers of water moving at different speeds may enable organisms to become dispersed over a wider area than would be the case if all members of the population drifted in the same mass of water. Furthermore, a species would tend to become divided into small, separate populations if all its members drifted at the same level, since they would not normally encounter any directional stimuli to horizontal migration. Gene flow is promoted

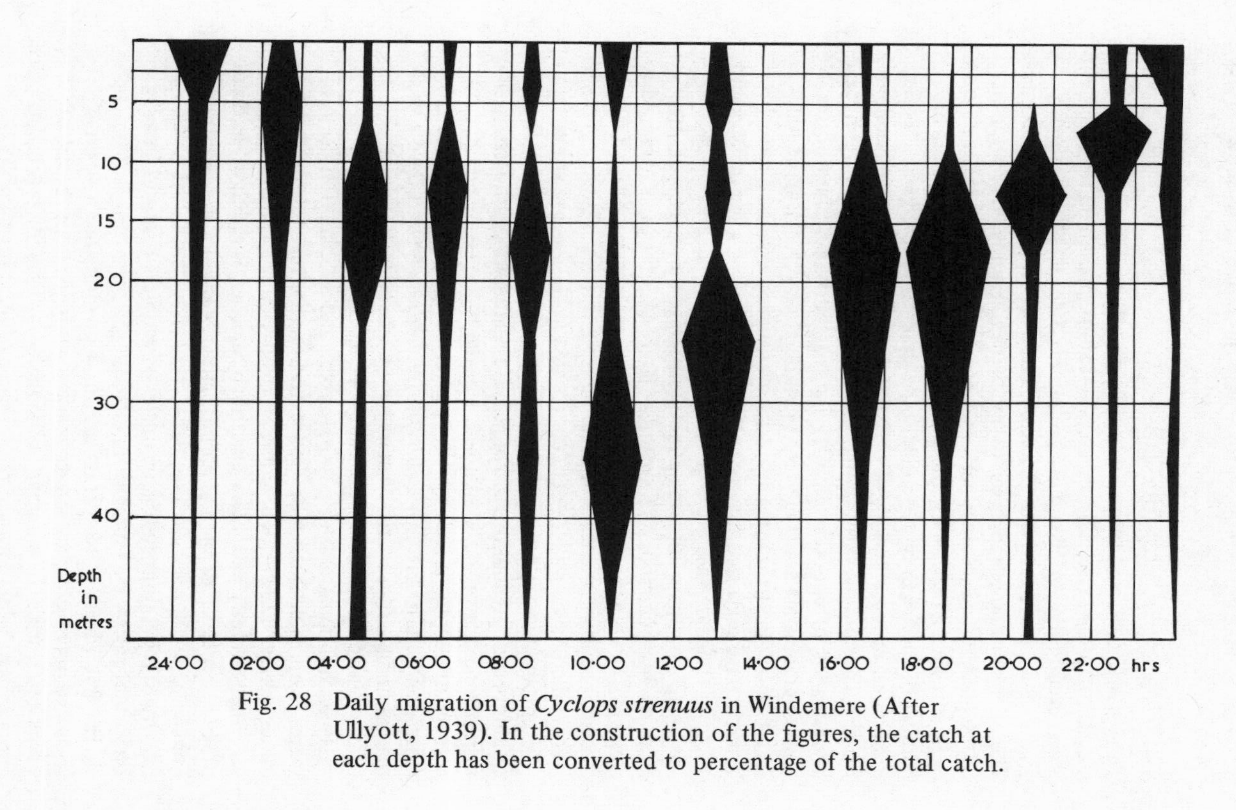

Fig. 28 Daily migration of *Cyclops strenuus* in Windemere (After Ullyott, 1939). In the construction of the figures, the catch at each depth has been converted to percentage of the total catch.

throughout these populations, however, by animals migrating at different times to different depths where the water layers show varying speeds of current flow.

Seasonal rhythms

The volume and constitution of both lake and river plankton vary according to the seasons. In temperate regions, maximum development is achieved in the spring and early summer. After this there comes a gradual decline, followed by a second, but smaller, autumnal maximum. Then there is a more rapid fall to a winter minimum. The composition of the plankton varies since each species of plant or animal has its own optimum season or seasons, and the coincidence of the maxima of a number of species in spring and autumn determines the general peaks in numbers at those periods of the year. Diatoms tend to predominate from late autumn to early spring, while Cyanophyceae, Chlorophyceae, *Peridinium* spp. and other dinoflagellates are mostly summer forms. No doubt numbers are controlled by a combination of environmental factors including light intensity, temperature and dissolved salts. Many workers have suggested that vernal and autumnal maxima in lakes with a thermocline, are causally correlated with disturbances at those seasons. Temporary full circulation then brings up reserves of nutrients from the hypolimnion. As Carpenter (1928) points out, however, this attractive hypothesis cannot provide a full explanation of the facts since both spring and autumn peaks are found in lakes, such as Lough Derg, which have no thermocline.

Reproductive rhythms are important in determining seasonal peaks of numbers in populations of zooplankton. Most kinds undergo periods of rest when active stages disappear and the species is maintained by eggs, or developing instars resting in diapause. Some species, such as *Leptodora kindti* (Fig. 26) and *Bythotrephes longimanus* have a single maximum in early summer: others have a second maximum at the time of the autumn peak in plankton numbers (Carpenter, 1928).

Peaks in populations of phytoplankton and zooplankton occur in major rivers such as the Nile. Concentrations are very low during the annual phase of high flood water, when adverse conditions of rapid flow and high turbidity exist. The subsequent increase is rapid, when there are high concentrations of plant nutrients. Reservoirs are important in enhancing the development of plankton (Talling and Rzóska, 1967). The hydrology, flora and fauna of the Nile are the subject of an important review by Hammerton (1972).

In some waters of temperate regions, a phenomenon known as 'seasonal polymorphism' is very marked among various species of plankton. During the summer, while rapid multiplication is taking place, there may be a temporary predominance of local forms while the original type reappears in the following winter (Fig. 29). Wesenberg-Lund (1911), who investigated these variations in Danish lakes, found that, in organisms as unrelated as species of *Ceratium* (Dinoflagellata) *Asplanchna* (Rotifera) and *Daphnia* Crustacea), there was a general manifestation of summer varieties with well-marked crests, elongated spines and other extensions of the body surface, compared with the more compact forms of winter generations. He suggested that these variations might be related to seasonal changes in the temperature of the water which would affect its density and viscosity. Reduced viscosity in summer would increase the difficulty of floating. This type of seasonal polymorphism is limited to shallow lakes with a wide annual range of temperature. In other cases variations may be irregular, or with a reduced development of spines in summer (Carpenter, 1928).

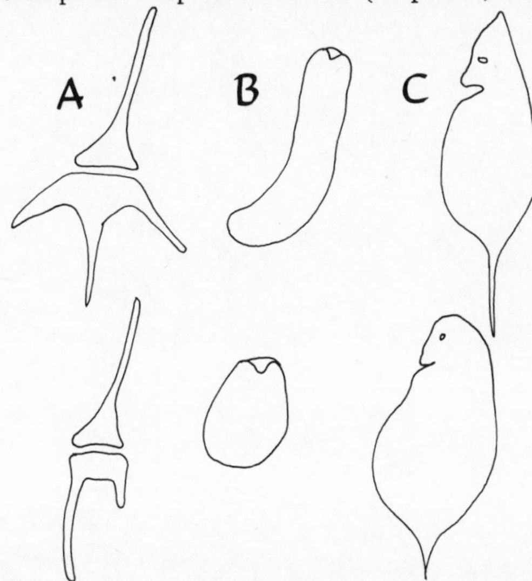

Fig. 29 Seasonal variation in some plankton types (After Wesenberg-Lund, 1911). (a) *Ceratium;* (b) *Asplanchna;* (c) *Daphnia.* Upper row: summer forms; lower row: winter forms from the same lake.

The periodicities of rotifers have been studied in great detail (Wesenberg-Lund, 1930), but can only in slight degree be identified with those of Cladocera. Two kinds of females are found in Rotifera, the young are rarely cared for and moulting does not take place. Consequently, yearly variations in the external medium have much slighter capacity for creating regular seasonal variations in these animals than in Crustacea; an internal cycle is thus more influential among rotifers.

In this chapter I have tried only to outline some of the more important environmental factors that influence flora and fauna of inland waters. I have not attempted to make a summary of the science of limnology, and the reader would be mistaken were he to regard the examples quoted as providing more than a very brief selection of the voluminous literature on the subject.

13 THE SELECTIVE INFLUENCE OF THE HABITAT

The various environments described in previous chapters can be classified according to their climates or by topographical features. Indeed, as Haviland (1926) has suggested, the topography of the country and the variations of the climate represent the same part in the make-up of the environment, figuratively speaking, as morphology and physiology play in the life of the organism.

Thus, irrespective of climate, the mammalian faunas of deserts, open savannah, steppes, tundra and high plateaux have certain features in common. In each, the dominant forms tend to be either large cursorial herbivores and their cursorial carnivorous followers, or subterranean rodents with social habits. The structural adaptations of limbs to running and leaping, or to digging, are often similar in unrelated forms from different parts of the world.

Again, the animals of tropical rain-forest, temperate deciduous forest, or of taiga, tend to show similar adaptations to arboreal life, but species adapted to climbing are naturally absent from open country. Nevertheless, where the plain or desert is broken by cliffs and ravines, it is frequently occupied by a sedentary population whose members show affinities with the inhabitants of forests elsewhere. Thus, some hyraxes *(Procavia* spp.*)* live in grass and desert regions; others *(Heterohyrax* spp.*)* live among rocks; but forest species of the genus *Dendrohyrax* inhabit hollow trees. Again, the European tree-creeper *(Certhia familiaris)* is an inhabitant of woodlands, parks and gardens with large trees. It nests behind loose bark, in split trees and behind ivy. The related wall-creeper *(Trichodroma muraria)* of central Europe inhabits rocky ravines, earth cliffs and ruins, breeding in deep crevices on inland cliff faces and among rocks. It lacks only the stiff supporting tail feathers of the tree-creeper.

In this chapter, we will first consider the adaptations of forest animals for climbing trees, and the cursorial adaptations of larger animals of open country. Attention will then be directed to the adaptations of plants and animals to life in one of the most extreme and inhospitable of terrestrial environments, the desert. Finally, we shall discuss the role of migration, hibernation and aestivation in enabling animals to exploit environments that are favourable for life during only a part of the year, and conclude with a review of the influence of the habitat on the coloration of its inhabitants.

Arboreal and Scansorial Adaptations

Many forest animals are adapted for arboreal life. Even primarily ground-dwelling species are able to ascend trees on occasion. Leopards, jaguars *(Panthera onca)*, small cats, pangolins, rodents and insectivores can climb well without being predominantly arboreal. The same applies to gorillas, although these are handicapped by their great weight and size. The majority of tree-dwellers are branch runners: they live and progress on all fours on the upper surface of the branches. These include lemurs, pangolins, opossums, squirrels, tree-kangaroos, chameleons and so on. Sloths (Brachypodidae), which move suspended beneath the branches by powerful recurved claws, only occur in Central and South America, but bats everywhere hang upside down by their hind limbs when at rest. Guereza monkeys *(Colobus polykomos)*, chimpanzees and gorillas are brachiators and swing from branch to branch and tree to tree by means of their fore limbs. African and Asian monkeys do not have prehensile tails like some of their relatives in the New World, but they are very much better climbers. Prehensile tails, like hook-shaped hands, tend to occur in relatively sluggish animals whose agility is not sufficient to enable them to recover easily from a false movement. Nevertheless, I once saw a Siamang gibbon *(Symphalangus syndactylus)* in Malaya fall a considerable distance before it managed to grab a branch and check itself.

Chameleons have prehensile tails, but they are extremely slow and placid creatures. The same is true to some extent of arboreal vipers. The vine or twig snake *(Thelotornis kirtlandii)* is actually a savannah species of South, Central and East Africa. It has a long whip-like body which it can make rod-like by tightening all its muscles in order to bridge a gap from one branch to another. The ventral scales of tree-dwelling snakes are usually stiffened by transverse keels which give added traction on rough bark surfaces (Cloudsley-Thompson, 1975e).

The feet of arboreal animals may be prehensile or non-prehensile. In the latter, the claws are usually well-developed, as in the squirrels and cats. The plantigrade feet of the North American tree-porcupines *(Erithizon* spp.*)* are armed with long, curved claws, in addition to which the soles bear spines and tubercles which aid in climbing. Prehensile hands and feet are modified for grasping by one or more of the digits being inserted so that it can be opposed to the other digits. Primates have an opposable thumb and, except in man, an opposable big toe as well. In anthropoid apes and most monkeys, the digit is flattened, which gives more gripping power, and the claws are replaced by nails. Opposable digits are also found in parrots, woodpeckers, chameleons and opossums. Among mammals, feet of the prehensile type are only found in primates and marsupials. In the latter they are associated with syndactyly, or fusion of the digits. This occurs in kangaroos, most of which are no longer arboreal. Even the Australian tree-kangaroos *(Dendrolagus* spp.*)* have to rely upon their claws and broadened soles for security. The loss of the hallux indicates a terrestrial ancestry before which, like all marsupials, their forebears must have been arboreal.

Adhesive pads are sometimes found, either on the tips of the digits, or on the soles of arboreal animals such as tree-frogs, geckoes, in which they function by friction rather than suction (Fig. 30), and tree-hyraxes *(Dendrohyrax* spp.*)*. Tree-frogs usually have well-developed adhesive discs on their digits. In climbing, the front legs are first raised and then the hind legs are pushed forwards, as when a human being climbs a tree. On smooth tree trunks, grass stalks or leaves, the frog presses its stomach against the support. This possibly adds to the adhesion and certainly increases friction. In balancing on a slim twig, the toes grip around the support, like those of a chameleon (Cloudsley-Thompson, 1975e).

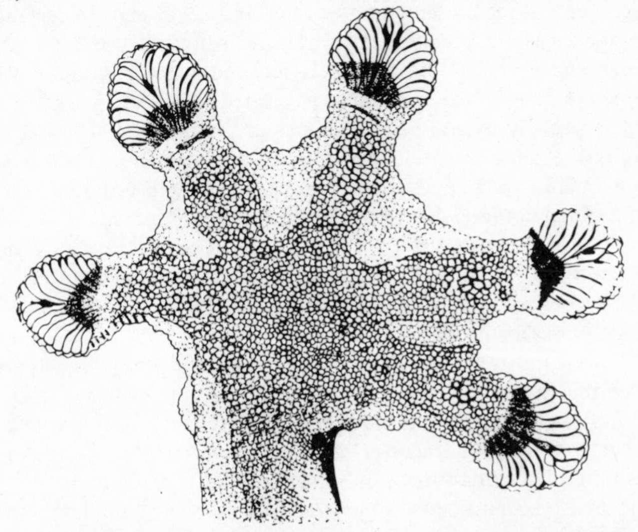

Fig. 30 Sole of foot of a gecko showing adhesive pads on the toes (From Cloudsley-Thompson, 1969). A horizontal force is produced by a pinching movement of the digits.

Although baboons are generally terrestrial inhabitants of wooded savannah, two related species, the mandrill *(Papio sphinx)* and the drill *(P. leucophaeus)* have become adapted to forest life, though they still retain their terrestrial feeding habits and live in troops. Prehensile tails are found in some Neotropical monkeys of the family Cebidae, which includes the spider-monkeys *(Ateles* spp.*)*, howlers *(Alouatta* spp.*)* and capuchins *(Cebus* spp.*)*. They are best developed in the spider-monkeys where they function as an extra hand. Perhaps, as a correlation, the real hands have lost the thumb, although the four remaining digits form an efficient hook for suspending the body. A prehensile tail is also found in the arboreal ant-eaters *(Tamandua* spp.*)* of the New World. Other small forest mammals include the tree pangolin *(Uromanis longicaudata)* and the attractive potto *(Perodicticus potto)*.

The suborder Anthropoidea is sharply divided into two series, the Old and New World Primates in which parallel evolution has taken place during the long period of South American isolation. The Platyrrhini are distinguished by the fact that the thumb is not opposable and, as already mentioned, sometimes reduced. The tail may be prehensile.

Adaptation to climbing implies a number of modifications to the body as well as to the hands and feet. The limbs tend to be elongated, especially their proximal segments, and the animals are generally plantigrade, walking on the palms of their hands and feet. In certain lemurs the tarsus is elongated, but this is probably because the creatures leap as well as climb. The clavicle and scapula bones are well developed in comparison with those of related terrestrial species. The clavicle, especially, would be a hindrance to the fore-and-aft swing of the limbs of a zebra or antelope but, in a climbing mammal whose arms are subjected to varied and violent strains, a clavicle is essential as it withstands the compression of the powerful breast muscles. The ileum or hip-bone is broadened as a support for the viscera in Primates, and even more so in American sloths (Brachypodidae) whose inverted posture necessitates additional support (Lull, 1940).

Many arboreal animals, in addition to birds and bats, have developed powers of volant or gliding flight. A fold of skin or patagium, supported between the limbs on each side of the body, acts as a plane. This is found in flying-squirrels and the flying-lemur *(Galeopithecus volans)* (Fig. 31). In the flying-dragon *(Draco volans)* and related species of lizards that inhabit the forests of South East Asia, the body is depressed and can be extended sideways by a pair of large wing-like membranes supported by five or six elongated ribs. Other flying reptiles include the flying geckoes *(Ptychozoon* spp.*)* of the same regions. These have been reported to glide through the air, buoyed up by the scaly fringes which run along the sides of the head, limbs, body and tail, and by the webs between the digits. Several so-called 'flying-snakes' have been recorded, such as *Chrysopelea* spp., which descend obliquely through the air with bodies rigid and ventral side concave. Volant Amphibia include the flying-frogs *(Rhacophorus* spp.*)* (Fig. 31) whose webbed feet sustain them in prolonged leaps from one tree to another. There are also rudiments of patagia in front of, and behind, the arms.

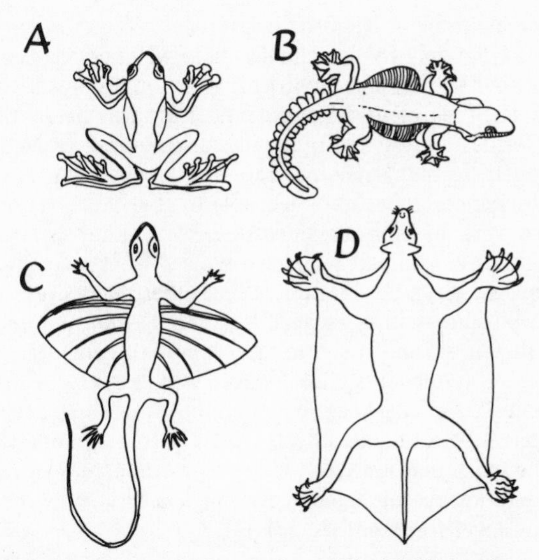

Fig. 31 Gliding vertebrates (After Lull, 1940). (a) Flying-frog *(Rhacophorus)*; (b) Flying gecko *(Ptychozoon)*; (c) Flying-dragon *(Draco volans)*; (d) Flying-lemur *(Galeopithecus volans)*.

The adaptation of animals to flight provides a very fine example of convergent evolution. Although only insects, birds and bats, like the extinct pterodactyls, are capable of sustained flight, several other animals can glide for considerable distances. These include flying-fishes, flying-frogs, flying-lizards, flying-squirrels and so on.

The requirements for flight in vertebrates are strict. Gravity must be overcome, so the flying animal must be comparatively light in relation to the strength of the muscles that move its wings. This condition is usually achieved by the evolution, in birds, of hollow bones with thin outer walls. The backbone must be strong and the pectoral muscles powerful: the breast bones are usually enlarged to provide for their attachment. The hind limbs are usually modified for landing. In addition, flying animals must have well-developed sense organs – eyes in birds or a sonar system in bats – a delicate sense of balance, and nervous control with consequent specialisation of the cerebellum.

Volant adaptations have been separately evolved at least thirteen times among mammals, but only the bats have attained the power of true flight. In these, patagia are extended between the body, the limb bones and the four elongated fingers. The pollux or thumb is free and clawed: it is used in crawling and climbing. The wing membrane of pterodactyls was supported by a little finger, which was enormously elongated, while the bird's wing consists of feathers attached to three digits which are more or less fused together.

Insects are the only flying animals whose wings appear to have evolved *de novo* and are not modifications of other, pre-existing structures. They represent developments of thoracic paranotal lobes. Some authorities suggest that flying insects may have evolved from small terrestrial forms carried by wind into the upper air. Those with the largest paranotal lobes were best able to glide in a controlled fashion and were therefore the most successful. It has also been suggested that flying insects may have evolved from roach-like, flightless ancestors. In Carboniferous times these were preyed on by mygalomorph spiders: they escaped by jumping from the tree-ferns on which they lived, and glided to safety on their paranotal lobes. When their prey took to the air, spiders evolved webs and trapped them in flight. Finally, it has also been suggested that wings may have evolved from the tergal flaps of male insects which wafted attractant scent towards the females. Primitively, the female mounted the male and, in many modern insects, the females are wingless while only the males are alate (Cloudsley-Thompson, 1975c).

Cursorial Adaptations

The larger animals of open country have evolved great speed and endurance. This is related to a number of factors: the absence of hiding places, the need to travel considerable distances to water, and to migrate. Even among birds, cursorial habits may be well-developed, as in the ostrich *(Struthio camelus)*, bustards of various genera, and the secretary bird *(Sagittarius serpentarius)*. Nevertheless, it is in the mammals that cursorial types are best developed.

Cursorial adaptations of vertebrate animals have been analysed in great detail by Gray (1968). Speed is achieved by increasing the length of the limb, and especially of its distal bones. For example, the limbs of horses and other equines are elongated distally by the formation of cannon bones from the third metatarsals. All toes, other than the third, have been lost and the humerus and femur are comparatively short. Concentration of muscles is thus produced at the proximal end of the limb. This lightens the limb so that it can be moved rapidly and, at the same time, increases their angles of insertion, giving higher propulsive components across the shaft of the femur. The muscles are thereby shortened and made thicker, which increases their power and speed of contraction.

In contrast, burrowing animals tend to have very short and powerful limbs. Shortening is particularly marked in the distal bones, which are unusually strong with prominent tuberosities for muscular attachment. Other adaptations include a tapering shape with narrow shoulders, a pointed snout, enlarge incisor teeth, fusion of the sacral vertebrae and a tendency towards fusion in the bones of the loin and neck (Lull, 1940).

Cursorial adaptations may be modified secondarily. The sitatunga, or marsh-buck *(Tragelaphus spekei)* of East Africa is well adapted to living in swamps. It has long, spreading hooves and hairless fetlocks. As it walks or runs on the tangled weeds, the two halves of the slender hooves spread widely and provide ample support for the body. The ancestors of the sitatunga were almost certainly speedy savannah antelopes that later took refuge from predatory enemies in tropical swamps, to which their descendants became adapted in the manner just described.

Size

Carl Bergmann's rule, first formulated in 1847, states that warm-blooded a nimals and birds from colder climates are larger, and therefore have proportionately less surface area from which to lose heat, than corresponding homeotherms of warmer regions (p. 160). (See discussion by Cloudsley-Thompson, 1975d). Although the African elephant *(Loxodonta africana)* may appear to be an exception, its surface area is increased approximately one third by its enormous ears, which have a rich blood supply and probably serve as radiators for cooling the body during the heat of the day.

The ecological significance of Bergmann's rule has, however, recently been challenged by McNab (1971) who points out that a positive correlation of weight with latitude in homeotherms cannot normally depend upon the physics of heat exchange, and that an animal does not live on a per-gram basis, but rather as an intact individual. Only the smallest species of a group of similar predators normally conforms to Bergmann's rule, and most of the widespread mammals in North America do not do so. McNab suggests, instead, that a correlation of body size with latitude in carnivores may reflect the size of the available prey. This, in turn, is influenced by the distribution both of the prey and of other predators with which it must be shared.

Desert Adaptation

The rigours of arid environments have engendered many striking similarities in the morphological and physiological adaptations of plants and animals (Hadley, 1972). Desert plants partly compensate for their lack of mobility by having an extended range at which their metabolic activities function. Their resistance to heat is due, in some degree, to increased molecular stability, particularly of enzymatic proteins whose bonding is strengthened by reduction in the degree of hydration. This phenomenon makes possible the growth of desert lichens on rock substrates whose temperature may exceed 70°C (Treshow, 1970). Likewise, the upper lethal temperatures of desert ectotherms such as arthropods (Cloudsley-Thompson, 1962d; 1964a) and reptiles (Cloudsley-Thompson, 1971a) are higher than those of species from other terrestrial biomes.

Important mechanisms for reducing and dissipating potential heat load include orientation, coloration and surface texture, surface structure and components, and hairs, spines or other surface projections. Orientation in relation to the sun determines the amount of radiation absorbed. Such behaviour is found among desert animals as dissimilar as locusts and camels. Vertically oriented pads and stems of desert cacti achieve the same effect (Treshow, 1970).

Incident radiation is sometimes reduced by projections and irregularities on the surface of desert organisms. The spines of desert cacti not only protect the plants from grazing animals, but also absorb and reflect much of the incoming solar radiation. The cholla *(Opuntia bigelovii)* may be covered with such dense spines that the surface of the stem is completely obscured and the temperature of the underlying tissues as much as 11°C cooler than they would be if the spines were absent (Pond, 1962). In addition, air is trapped against the surface of the stem, creating a boundary layer which minimises heat loss and heat gain by convection.

Similar boundary layers are created in animals by hairs, scales, feathers and fur (Schmidt-Nielsen, 1964). Thus, a steep gradient of temperature across the boundary layer trapped by the fur of desert mammals reduces the flow of heat from the environment. In the case of arthropods, scales and hairs likewise delay the rate at which body temperatures increase when exposed to sunlight (Hadley, 1972).

The subelytral cavity of desert Tenebrionidae not only serves to reduce water loss by transpiration through the spiracles (Cloudsley-Thompson, 1964b; Dizer, 1955) but it also serves to insulate the body tissues from the hot cuticle above (Hadley, 1970).

The size of the body is an important factor in determining the rate of movement of heat between an organism and its environment (p. 147). This relationship between area and volume is particularly important in desert animals, both in relation to thermal balance and to water conservation. Larger desert animals, such as camels *(Camelus dromedarius)* and ostriches *(Struthio camelus),* can tolerate high body temperatures and thus economise in water used for cooling and in energy. The elevated body temperatures narrow the gradient between the ambient temperatures and those of the body, thereby reducing the rate at which heat is gained by the environment. This is one reason why sweating, compared with panting, is an inefficient method of cooling the body in relation to the amount of water used in transpiration.

Desert plants and animals, both invertebrate (Cloudsley-Thompson, 1964a; Cloudsley-Thompson and Chadwick, 1964 etc.) and vertebrate (Schmidt-Nielsen, 1964 etc.) show many morphological and physiological adaptations for the conservation of water; and the analogies between plants and animals in this connection have been outlined by Hadley (1972).

Migration

Unlike plants, animals are able to move rapidly from one place to another and thus exploit, for short periods, habitats that are not favourable for their survival throughout the entire year. For example, many birds breed in Arctic regions during the summer, when the days are long and there is an abundance of insect life (p. 141) on which to feed their young. Before winter returns, however, they migrate to warmer climes stimulated by the proximate factor of decreasing photoperiod (pp. 39;125). Migratory behaviour occurs in various phyla of the animal kingdom – among terrestrial forms it is especially conspicuous in Arthropoda and birds.

Amphibians migrate to fresh water to breed, and seasonal migrations also occur in reptiles. There are numerous instances of migration for breeding purposes among mammals. Caribou *(Rangifer tarandus)* will travel 800 km (400 miles) from winter retreats to their pairing grounds, and mule-deer *(Oedocoileus hemionus)* may cover as much as 240 km (150 miles). Regular migratory movements, associated with seasonal reproduction, have been described in many species of African game animals (Bouliere and Hadley, 1970; Cloudsley-Thompson, 1967b) and similar migrations have been reported from other continents.

Birds

The migration of birds is the subject of many articles, chapters and books. C. Linnaeus wrote an academic thesis on the subject in 1757. More recent accounts have been published by Dorst (1962), Rowan (1931), Thomson (1926; 1949) and others. The existence of the phenomenon has been known since ancient times for Homer (twelfth) century B.C.) wrote:

> So when inclement winters vex the plain,
> With piercing frosts, or thick descending rain,
> To warmer seas the Cranes embodied fly,
> With noise and order through the midway sky.
> (*The Iliad,* Bk. III, Trans. A. Pope, London, 1717.)

In the Book of Job (XXXIX, v.26) we find the well-known passage: 'Doth the stork fly by Thy wisdom, and stretch her wings towards the south', and in Jeremiah (VIII,v.7) the prophet declares 'The stork in heaven knoweth her appointed times; and the turtle-dove, and the crane, and the swallow, observe the time of their coming'. Anacreon (5th century B.C.) welcomed the return of the swallows in spring, and was correct in his assumption that Egypt was one of their winter retreats (Cloudsley-Thompson, 1960a).

Very many species of birds are to some extent migratory and, although migration reaches its greatest development among species breeding in higher latitudes, migratory forms include several that are wholly confined to the tropics. At the same time certain Arctic species are sedentary. Migration is, nevertheless, related to climate in that it is a form of behaviour adapted to seasonal changes in the environment. In its more extensive forms it involves a change in latitude, although the direction of movement is often modified by circumstances of climate and geography. In some instances there are marked differences between populations of a single species, or even between individuals or age-groups within a single population, as to the time, extent and route of migration (Thomson, 1949).

Birds have spread over the world from different centres for varying distances and at different times; but, of course, long after Continental Drift took place, so this cannot have been a factor in the evolution of migratory vehaviour. Nevertheless, geographical features do affect migration and birds often follow the coastline, river valleys and so on (Darlington, 1957).

In general, migratory species tend to breed in the colder part of their range, and during the summer when there is abundant food and long hours of daylight. Of migratory birds living entirely in the tropics, the majority breeds during the rains although some reproduce in the dry season. Thus, the Abdim's stork *(Sphenorhynchus abdimii)* nests during the rains in the northern Guinea savannah and southern Sudan transition zone (p. 59) where it is regarded as a harbinger of rain. It moves south in October and November, passing through the East African grass belts during the rains there and, during the dry season, inhabits the wet grassland from Tanzania to the Transvaal during the southern rainy season. In contrast, the pennant-winged nightjar *(Cosmotornis vexillarius)* breeds in the southern spring just before the rains and then migrates to the north of the equator (Cloudsley-Thompson, 1969).

Migratory species are nearly always insectivorous, both in temperate and in tropical regions (Moreau, 1966). No doubt considerations of food ultimately determine all such movements – either directly, because the breeding area later becomes ecologically unfavourable, or because the wintering area is unsuitable for breeding. In general, it appears that migration is a more important factor in bird ecology north of the equator than south of it.

The climate of the world has fluctuated considerably since migratory habits first evolved. The increasing aridity of the Sahara, during the last 5,000 years, will greatly have increased the length of their migratory journeys. This may means that the whole pattern of trans-Saharan migration has changed during this period, and that birds have become adapted to accumulating reserves of fat before flying across the desert (Cloudsley-Thompson, 1969). It is paradoxical that the fattening process should take place during winter and the early spring when the savannah and desert are extremely dry and insect life is comparatively scarce.

The recurrent proximate stimuli that evoke migratory behaviour in birds are of two kinds. Endogenous factors influence the annual physiological cycle, while exogenous factors, especially photoperiod, release the activity (Rowan, 1932). Physiological readiness is linked with the changing condition of the gonads, and is influenced by hormones. But the relationships are by no means simple and different releasers may operate in different species of bird.

Arthropods

The evolutionary significance of migration has often been disputed because, as Elton (1927; 1930) points out, a large number of individuals must fail to find a suitable habitat and so die without leaving progeny. This applies more to emigration (Heape, 1931) than to true migration. I have argued elsewhere that the selective advantage to a species in evolving emigratory behaviour may be explained in terms of group selection (Cloudsley-Thompson, 1961; 1971b) although, perhaps, not quite in the form envisaged by Wynne-Edwards (1962).

Insect migration has produced many physiological and ethological problems analogous with those posed by birds and other groups of animals. Williams (1958) suggested that it might function either as an overflow from an over-populated area, an escape from overdevelopment of parasites and predators, or as an escape from seasonally recurring food shortages or other unfavourable conditions. Although these possibilities have not been supported by the available evidence, Williams makes an important point in noting that Heape's (1931) categories of migration, emigration and nomadism can no longer be regarded as valid. Insect migration may be defined as continued movement in a more or less definite direction, and consists of three phases: (a) emigration from the original home, (b) transmigration, (c) immigration. Among birds, which are relatively long-lived creatures, all three take place within the lifetime of an individual but, in insects, the migratory flight is halted when reproduction takes place, and is then continued by the offspring. For instance, the monarch butterfly *(Danaus plexippus)* extends its range in summer to Canada and the Hudson Bay. In autumn it moves southwards to Florida, California and Mexico, where the winter is passed in semi-hibernation. Northward migration begins in March, and the migrants arrive in southern Canada at the beginning of June. Here they breed, passing up to three generations before the return journey begins. It is significant that the individuals which migrate south in autumn are the same as those which fly north again in spring (Urquhart, 1960).

The migrations of butterflies appear to be stimulated by ecological events at the point of origin more directly than are bird migrations. A likely cause of mass migration might be the local scarcity of a particular food-plant which, in turn, could be determined by drought or other climatic factors. Neither are the proximate factors stimulating migration known with any certainty, nor what determines the direction taken. Migration may be stimulated by changes in day length and temperature or, more directly, by the local failure of a particular food. In the tropics, the alternation of wet and dry seasons may provide stimulation and, in this connection, it is significant that most migratory species are inhabitants of the savannah which is, of course, seasonally much more varied than forest (Owen, 1971).

In an important review of migration in terrestrial arthropods, Southwood (1962) suggests that the prime advantage of migratory movement lies in its enabling a species to keep pace with changes in the location of its habitats. If Southwood's hypothesis is true, the level of migratory movement in any species should be geared to the rate of change of its habitat and be highest in those species with the most temporary habitats. When the migratory movements in the major taxa of terrestrial Arthropoda are considered in this light, it is found that the level of migratory movement is positively correlated with the impermanence of the habitat. Southwood (1962) therefore concludes that the changing pattern of the environment has been the primary factor in the evolution of migratory movement in arthropods. He agrees, however, that migration also plays an important part in the population dynamics of the denizens of temporary habitats, being on the one hand a defence against predators and, on the other, a means of reducing density and preventing overcrowding.

More recently, Baker (1968) has pointed out that none of the explanations mentioned above can account for a flight that has a predominant geographical direction, a feature which, for many years, has been considered to be characteristic of butterfly migration (Williams, 1930). He suggested, however, that during their evolution many species of butterflies have become adapted to larvel foods that occur in small localities whose distribution changes constantly. These butterflies fly from one site to another, maintaining a constant angle to the rays of the sun and this angle varies with the ambient temperature. In a later paper (Baker, 1969) this idea is followed up in greater detail and it is shown that the peak of direction taken by the first and second broods of multi-brood species is towards a lower temperature. It is a function of the geographical temperature gradient experienced during the larval development of the offspring. In autumn, there is a major change in the direction of peak flight towards a warmer temperature.

Physiological and ecological observations on a variety of species are consistent with the following generalisations about insect migration: (a) During migration, locomotion is enhanced while feeding and reproduction are reduced. (b) Migration usually occurs before the insect reproduces. It takes place, therefore, when reproductive values and colonising abilities are near their maximum. (c) Migrants usually rest in temporary habitats. (d) Both physiological and ecological parameters of migration are modified by environmental factors. Taken together, these criteria establish a basic strategy which can be modified to suit the ecological requirements of individual species (Dingle, 1972).

The migration of insects is a vast subject and, in the case of locusts, one of immense economic importance. It seems unlikely that it can result from the same combinations of physiological and environmental stimuli in different species or even within different populations of the same species. For further consideration of the phenomenon, the reader is referred to Southwood's (1962) review and the books by Williams (1930; 1958). In this chapter it has been discussed only in so far as it provides an example of one way in which inconstant environments affect their faunas.

Hibernation and Aestivation

Migration is not the only method by which animals overcome the problem of survival during seasons of unfavourable weather. An alternative, found in less mobile animals, is to hibernate or aestivate, as the case may be, in a state of diapause. The physiology of hibernation and aestivation in homeotherms have been studied intensively (Folk, 1966; Hoffman, 1964).

Their physiological responses to proximate stimuli are somewhat similar to those concerned with migration – there are similar metabolic adjustments, a reduction in endocrine activity and an increase in the deposition of fat. By permitting the body temperature to follow passively the ambient temperature, from 2° to 32°C, without this eliciting arousal, small animals and birds are able drastically to conserve their supplies of food and water. Animals that have adequate food accessible or in stores do not need to hibernate or aestivate. But species without external food reserves are compelled to fast, making use of internal reserves which need to be conserved as far as possible.

Hibernation is a comparatively rare phenomenon in mammals and is found only in animals the size of marmots or smaller. Of about 23 non-migrant species of this size which inhabit northern Canada, only four are known to hibernate (Bert and Grossenheider, 1952). Others may employ physiological mechanisms to combat an extreme environment or, like beavers *(Castor canadensis)* use engineering methods to maintain a safe environment in winter. Hibernation is little known in birds, probably because they are more mobile than mammals and therefore migrate. Some species, however, such as the white-throated swift *(Aeronautes sexatilis)* have the capacity to survive inclement conditions in a state of reduced animation and body temperature – usually for relatively short periods (McAtee, 1947), but for up to three months in the poor-will *(Phalaenoptilus nuttallii)* of western North America (Jaeger, 1948). All reptiles and amphibians of temperate regions spend the winter in suspended animation, and diapause is characteristic of Arthropoda and other invertebrates.

The physiology of aestivation in desert mammals has been studied less than hibernation, but is comparatively well understood in ground squirrels *(Citellus* spp.*)* and smaller rodents (Hudson and Bartholomew, 1964). Periodic dormancy, in which animals become torpid at low or high ambient temperatures, is probably polyphyletic in origin, having been evolved independently in several different groups of mammals, but the physiological mechanisms concerned are very similar. Aestivation in the dry season is characteristic of amphibians of the savannah and arid lands (Bouliere and Hadley, 1970): summer dormancy is likewise found in plants (Koller, 1969). The physiology of diapause in arthropods has been the subject of intensive study (Lees, 1955). It may be either obligatory and inherent; or facultative, in which case it can be induced by a number of different stimuli including low temperature, photoperiod or particular kinds of food (Andrewartha, 1952). Recent work on the subject has been reviwed by Beck (1968). Not only are stages in diapause temporarily independent of food and moisture, but they show enhanced resistance to drought, low or high temperature, and other adverse climatic factors.

Cryptobiosis

No account of resting stages and the suspension of metabolism in the lives of living organisms would be adequate without discussion of cryptobiosis, defined by Keilin (1959) as ". . . . .the state of an organism when it shows no visible signs of life and when its metabolic activity becomes hardly measurable, or comes reversibly to a standstill". Hinton (1960, 1971), however, restricts the term to describe a state in which metabolic activity is totally absent. Examples are afforded by bacteria, fungal spores and various seeds. They are reviewed by Hinton (1971). One of the most spectacular concerns seeds of the Arctic lupin *(Lupinus arcticus)* which germinated successfully after being buried deeply in Pleistocene silt which had been permanently frozen for at least 10,000 years (Porsild, Harington and Mulligen, 1967). The successful maintenance of the cryptobiotic state does not require interactions between cells or their constituents. It can be induced by freezing or by dehydration.

Although cryptobiosis by dehydration at developmental temperatures can be induced in many insect tissues, dehydration of the whole animal is now known only in the larvae of three species of flies, of which *Polypedilum vanderplanki* has been studied in detail by Hinton (1951; 1960; 1971). This midge breeds in small pools formed in shallow hollows in unshaded rocks in Uganda and Northern Nigeria. During the rainy season, these rocks may fill and dry several times; but the larvae are well adapted to this unstable environment because they can absorb water and dry up many times without harm. They survive in the dehydrated state for several years and, while desiccated, are able to tolerate extremely high or low temperatures. As a result of his research on cryptobiosis, Hinton (1968) has postulated that the first living systems may have originated abiogenically on land or in small pools that dried out periodically so that chemicals would have become progressively more concentrated in them.

Animal coloration

The adaptive functions of animal colours are the attainment of concealment for defensive and offensive purposes, of disguise or advertisement (Cott, 1940). The principles by which these functions are achieved may be fundamental but the results vary according to the environment. The predominant colour in forest is green; in desert, brown; in snowlands, white. Consequently, general colour resemblance will be achieved by green animals in forest, brown in desert, and white animals in snowlands. Where there are marked seasonal changes in the environment, seasonal dimorphism appears in the adaptive colours of the fauna. Defences against visually hunting predators, with especial reference to invertebrates, have been discussed in a thoughtful paper by Robinson (1969).

Forest

Almost every aspect of adaptive coloration is seen at its best in rain-forest. An environment which consists so largely of leaves, bark and stems has, naturally, engendered the evolution of disguises by which animals tend to resemble leaves, bark or twigs. Resemblances to leaves appear in many groups of animals, including cockroaches, leaf-insects, praying-mantids, grasshoppers, caterpillars, butterflies, moths and even toads *(Bufo typhonius)* (Cott, 1940).

In addition to cryptic forms, many brilliantly coloured species of animals inhabit tropical rain-forests. Their distribution depends to some extent upon light intensity. Beebe, Hartley and Howes (1917) tabulated the species of bird that they found in one area of forest in Guyana. Then, reckoning the gradations in light intensity from the dimness of the forest floor to the full sunlight of the tree-tops as between 1 and 10, the following assessments were obtained:

Estimated degree of light	1	3	5	10
Percentage of bright birds	0	8	50	83

Although the most brilliantly coloured species live in the brightest light, this does not mean that they are the most conspicuous. Indeed, among the flowers of the luxuriant vegetation of the upper storeys a drab bird might show up in contrast to its surroundings.

The rule that brightly coloured animals frequent light places apparently does not always apply to insects. Some of the most brilliant of the Hemiptera – yellow, red, black or metallic green in colour – live in deep shade; while species characteristic of clearings are often obliteratively coloured green or brown. Again, conspicuous butterflies of the sub-families Ithomiinae and Heliconiinae (Nymphalidae) inhabit shady places. But many shade-dwelling Lepidoptera are extremely inconspicuous. In some cases, as in certain Satyrinae, Ithomiinae and Riodinidae, the wings are denuded of scales so that they are quite transparent. In others, there is a full covering of scales but these are obliteratively coloured (Haviland, 1926). No doubt there can be only the broadest correlation between light intensity and percentage of cryptic species in the faunas of various forest habitats.

Various inhabitants of forest and wooded savannah in tropical and sub-tropical countries have their bodies decorated with white spots. Examples are afforded by the African bush buck *(Tragelaphus scriptus)*, the axis deer *(Axis axis)* of India, and the spotted cavy *(Cuniculus paca)* in South America. Where shafts of sunlight burst through the canopy, such markings distributed on an animal's back and flanks produce an extremely cryptic effect.

In temperate latitudes, where the leaves fall in autumn, spots would be conspicuous during the winter months. It is significant, therefore, that species such as the fallow deer *(Dama dama)* and the Japanese deer *(Sika nippon)*, which inhabit deciduous forest, undergo a seasonal colour change. In summer, they are spotted but, during winter, they assume a uniform grey or brown colour (Cott, 1940).

The features of adaptive coloration seen in rain-forest are also found, though to a lesser degree, in deciduous forest and taiga. Adaptive coloration must also have important survival value in grasslands and desert where good cover is lacking. While cryptic coloration is the general rule in such environments, there are striking exceptions.

Grassland

It is not by chance that the largest herbivores and swiftest predators should inhabit the grasslands of the world for concealment is difficult in open country so that size, speed and social behaviour are often the only defence. Such adaptive coloration as is found is probably effective mainly in trees or long grass – situations in which an animal is particularly vulnerable to predation. For example, the zebra *(Equus burchelli)* is extremely conspicuous in full sunlight and in open country but, where there is thin cover and at dusk, when he is most liable to be attacked, he is one of the least easily recognised game animals (Cott, 1940; Cowles, 1959). As I have pointed out elsewhere (Cloudsley-Thompson, 1967b), however, while the markings of the zebra are very obvious at close range, from a distance the black and white stripes merge together to form a light grey colour that is actually slightly less conspicuous than the dark colours of topi, tiang or wildebeest.

In countries with well-marked wet and dry seasons, insects with aposematic colouring are relatively less abundant during the dry period when competition is greater (Poulton, 1908).

Desert

Most of the inhabitants of deserts are either black or pale in colour, resembling their background. In many desert mammals the under-surface is very pale or quite white. Desert species and subspecies differ from their near relatives from other environments just as much in their pale ventral surface as in their buff or sandy backs. Moreover, the pale ventral area is often extended over the flanks in desert forms to a greater extent than it is in related species from other habitats. The same phenomenon occurs in desert spiders, centipedes, woodlice, insects, lizards, snakes and birds. Furthermore, desert-dwelling animals are not coloured fawn, brown, cream or grey indiscriminately. There is often a very close similarity between the creature and the soil of the particular type of desert on which it is living, although the relationship may not always be apparent (Benson, 1933; Bodenheimer, 1957; Buxton, 1923; Sumner, 1921). Opinions are sharply divided as to the significance of desert coloration. Whereas Cott and I regard it as an adaptation to predation, Buxton and Bodenheimer favour a physiological interpretation. Hamilton (1973) ascribes a thermal function to melanism but agrees that desert colours are cryptic. Birds with desert coloration tend to be sedentary and, in the Namib desert, occur as endemic species or subspecies (Willoughby, 1969. (See p. 105).

In addition to desert coloration, blacks and whites are also common. The occurrence of blackness in widely separated groups of animals is remarkable. Examples include ravens, wheatears, tenebrionid and scarab beetles, bees, flies and so on. Black is conspicuous and can be considered as a warning coloration in distasteful or poisonous animals. Its prevalence among desert animals may result from Müllerian mimicry, in which a number of different species, all possessing aposematic or warning attributes, resemble one another and so become more easily recognised. Numerical losses are reduced in teaching predators to avoid a common warning colour and the adoption of a common advertisement simplifies recognition. In Batesian mimicry, on the other hand, a relatively scarce, palatable and unprotected species, such as a bee-fly (Bombyliidae) resembles an abundant, relatively unpalatable or well-protected species such as a bee and, on account of its disguise, is ignored by potential enemies. In nature, Batesian and Müllerian mimicry tend to emerge into one another as model and mimic become relatively more or less distasteful.

The struggle for existence is especially severe in desert regions. Poisonous species are said to be more dangerous than their relatives from less arid environments, speedy animals faster, senses more acute. In a similar way, natural selection has played unusually heavily on the colours of desert animals. Cryptic species are exceptionally inconspicuous and Müllerian mimicry extremely common. Colours without marked adaptive significance are seldom, if ever, found (Cloudsley-Thompson and Chadwick, 1964).

The most striking support for the hypothesis that the adaptations referred to here are, indeed, correlated with the desert environment is afforded by the extraordinarily close resemblance between unrelated animals occupying similar ecological niches in different areas of the world. Examples are afforded by the kangaroo-rats *(Dipodomys* spp.) of America and the jerboas *(Jaculus* and other genera) of the Old World deserts; the hack-rabbits of North America, the hares of the Great Palaearctic desert and the marsupial quokka *(Setonix brachyurus)* of Australia. The American kit-fox *(Vulpes velox)* is superficially very much like the Saharan fennec *(Fennecus zerda);* road-runners *(Geococyx* spp.) of the New World, which are related to cuckoos, look like the coursers and pratincoles of the Old World which belong to the family Glareolidae. There are marked parallels between the iguanid desert lizards of America and the Palaearctic Lacertidae while the desert rattlesnake *(Crotalus cerastes)* of Arizona, apart from the possession of its rattle, is almost indistinguishable from the horned viper *(Cerastes cerastes)* (Fig. 18) of the Great Palaearctic desert (Cloudsley-Thompson, 1972a).

Snowlands

The birds and mammals of snowlands are usually in colour, often with darker markings. There are, however, species such as the elk *(Alces alces)* and musk ox *(Ovibos moschatus)* (Fig. 22) which do not become white during any part of the year, even in the coldest parts of their range. In certain species, such as the variable hares *(Lepus americanus* and other species), Arctic hares *(L. arcticus)*, ermine *(Mustela erminea)*, Arctic fox *(Alopex lagopus)*, caribou *(Rangifer tarandus)*, willow grouse *(Lagopus lagopus)* and ptarmigan *(Lagopus mutus)*, white colour is assumed only in winter and in the colder part of the range. Others, such as the polar bear *(Thalarctos maritimus)*, polar hare *(Lepus arcticus)*, snowy owl *(Nyctea scandiaca)* and Greenland falcon *(Falco candicans)* are white all the year round (Cott, 1940). The physiological basis of seasonal colour change among Arctic animals is still little understood, but there may be some evidence that lowering temperatures provide a proximate stimulus. (p. 148).

In contrast to mammals and birds, nearly all the invertebrate animals of the snows are black, or very dark in colour. This enables them to absorb the scanty warmth of the sun's rays upon which their metabolism depends. Dead insects, blown on to glaciers, are often to be seen in depressions where the radiant heat they have absorbed from the sun has melted for them an icy grave (Cloudsley-Thompson, 1965d). The 'snow-flea' *(Isotoma nivalis)* of Spitzbergen, like the larger Alpine *I. saltens*, forms great black masses sometimes extending over an area of about 900 sq cm or more, and a great black patch of earthworms of the family Enchytraeidae, over 400 m long, was found in snow at a height of 1,585 m (5,200 ft.) in Oregon.

Since Cott (1940) produced his encyclopaedic review of adaptive coloration in animals, increasingly sophisticated techniques have been used by ecologists and ethologists to re-examine the subject, and experimental methods have been employed to test hypotheses about Batesian mimicry, camouflage and startle displays (Blest, 1966; Rettenmeyer, 1970; Tinbergen, 1963). A particularly important series of papers has been published by Brower and his co-workers (for references, see Brower, Cook and Croze, 1967; Platt and Brower, 1968). Robinson (1969) emphasises four major points which, given further study, could yield data of considerable interest. In particular, the problem of recognition of the prey must provide an explanation of the fact that some Batesian mimics have achieved nearly perfect resemblance to their models: and that many mimics of sticks and leaves have evolved similarly complex and detailed resemblances. Only some species have done this, however. Others have apparently remained uninfluenced in this way by selection pressure. Again, the vision of insect predators may extend into the ultra-violet while that of owls lies at the infra-red end of the spectrum. Adaptive coloration must surely be related to the vision of the predator and, if there are more than one, presumably reflects a compromise between the visual powers of both.

14 THE INFLUENCE OF THE ORGANISM ON ITS HABITAT

Living organisms began to influence the terrestrial environment from the moment that the first algal spore managed to survive on land, or the first invertebrate wormed its way out of the ocean. We have already seen (Chapter 2) that soils are formed by mechanical, chemical and biological weathering of the parent rock. The L and A_0 layers are composed almost entirely of organic matter, while the A_1 layer has a relatively high organic content.

Recolonisation of Krakatoa

The way in which a sterile environment becomes colonised by plants and animals has been investigated both experimentally and through observing the recolonisation of land devastated by volcanic activity, floods and other natural disasters. Perhaps the most dramatic of such natural experiments within recent years has been provided by Krakatoa, 40 km east of Java. This was the scene of a tremendous volcanic explosion on 26 August 1883 which destroyed part of the island and covered the remainder so thickly with pumice and ash that most, if not all, living organisms were exterminated. After only three years, however, the soil had become thickly covered with blue-green algae (Cyanophyceae), while eleven species of ferns and fifteen species of flowering plants had established themselves. By 1897, there were twelve species of ferns and 50 of flowering plants; in 1906 there were 114 plant species and the composition of the flora had changed considerable.

Plants were promptly followed by animals in their colonisation of the island. As early as 1889, many spiders, bugs, flies, beetles, Lepidoptera and even a species of monitor-lizard *(Varanus salvator)* had arrived. In 1908 (25 years after the eruption) 263 species were collected, including 240 arthropods, four molluscs, two reptiles and sixteen birds. Investigations during 1921-2 yielded 573 species of animals including three of mammals, 26 birds and one snake *(Python reticulatus)*. Comparison with neighbouring islands showed that the fauna then had about 60 per cent of the expected number of species (Dammermann, 1922). As Hesse, Allee and Schmidt (1952) point out, about 90 per cent of the plants and animals present could have been brought by wind; rats and lizards were probably transported on driftwood or boats, crocodiles and pythons could have swum. The nearest island not destroyed by the eruption is 18.5 km away so the new inhabitants must have travelled at least that distance.

The consensus of biological opinion now favours the effectiveness of long-distance dispersal as a means of populating islands. The principles involved have been reviewed by Carlquist (1966) who points out that features exhibited by the biotas of islands are adaptive radiation, flightlessness in animals, loss of dispersal mechanisms in plants, and the development of new ecological habits and growth forms. Each of these is governed by a wide variety of factors.

Vegetation

Not only is vegetation important in the creation of soil, but it makes the land habitable for animals, creating microenvironments suitable for small forms and providing food and shelter for larger species. One of the more important influences of vegetation lies in mollifying the effects of climate. This is clearly seen where overgrazing or greedy agriculture has created desert of dust-bowl conditions. The ground is then no longer blanketed by plant cover. Insolation engenders high surface temperatures, rainfall tends to run rapidly away, causing erosion in the process, while the top soil is speedily removed in the absence of shelter from the wind. The difference between microclimatic conditions produced by plants and the general climate becomes increasingly important if the latter is less favourable. The vegetation of Arctic and Alpine regions utilizes solar radiation to produce a microclimate suitable for other plants and for many small animals, especially insects (Geiger, 1950).

The question whether vegetation, and particularly trees, can actually influence the climate has been much debated. Stebbing (1954) has argued strongly that the rainfall dries up when forest is destroyed by man. This results in still further degradation of the soil and vegetation. There appears to be no direct evidence, however, for climatic change resulting from deforestation, although it is obvious that a covering of vegetation retains moisture, prevents erosion, and shades the surface of the soil from excessive insolation.

Likewise, it is commonly believed that large bodies of inland water or swampy vegetation may influence climate and engender increased rainfall locally. Observations in Africa near Lake Rukwa, however, have shown that the effect is much smaller and more localised than is commonly believed. Even as far as 200 – 300 m from the shore, the relative humidity was still only 40 – 50 per cent. The humidity of the air over wet mud was as low as that of the air above dried up and burnt grass hundreds of metres inland, for there are strong vertical gradients in temperature and relative humidity near ground level (Michelmore, 1947). I have measured relative humidities as low as 40 per cent between 10.00 – 15.00 hrs in the middle of the Nile Sudd in December.

when the air temperature went up to 40°C (104°F) (Cloudsley-Thompson, 1969).

Fauna

The most important way in which animals influence their habitats is through overgrazing and destroying the vegetation. This is most marked in arid and semi-arid regions with intermittent rainfall, but examples also occur in other areas of the world. I have recently argued that, during the late Pleistocene, the Sahara experienced both pluvial and inter-pluvial periods when the climate was successively wetter and drier than it is today. At the end of the Pleistocene, the regions bordering its southern edge were richly supplied with lakes and rivers. These dried up between 7,000 and 3,500 years ago as the climate became progressively drier and warmer. Subsequent impoverishment of the flora and fauna has been due almost entirely to human activities – bad agriculture, felling of trees for fuel, and overgrazing by domestic stock, especially goats (Cloudsley-Thompson, 1971c).

The region of Chihuahuan semi-desert by the Sacramento mountains of New Mexico (Plate 2b) was covered with lush grass at the beginning of this century, but overgrazing by cattle has reduced it to its present impoverished state (Cloudsley-Thompson, 1975b). In temperate climates, forest can be changed into grassland by grazing alone, although in agricultural lands the process is usually accelerated by sowing with grass seed (Tansley, 1939). Wild animals tend to be migratory or nomadic when left to their own devices, and move long before they have caused too much damage to the environment. In many of the cases where they have caused damage, man has in some way interfered with the natural order. For example, introduction of red deer *(Cervus elaphus)* into New Zealand where they had no natural enemies made a profound impact upon the native forests, resulting in soil erosion on many watersheds. Until the beginning of this century, the population of white-tailed deer *(Odocoileus virginianus)* on the Kaibab plateau of Arizona probably numbered about 4,000. This was well below the carrying capacity of the range, but numbers were kept down by predation from pumas *(Felis cougar)* and wolves *(Canis lupus)*. Between 1907 and 1923, however, nearly 700 pumas and eleven wolves were shot, as a result of which the deer population increased very rapidly until it far exceeded its winter food supply. From a peak of 100,000, the herd declined to 40,000 in the winter of 1924-25. By 1931, it had dropped to 30,000 and continued to decrease until it fell to 10,000 ten years later. An optimum carrying capacity for the Kaibab plateau has been estimated at about 30,000 deer and this would not have damaged the vegetation (Leopold, 1943).

This is the story that has found its way into most general ecology texts, but it is not really supported by the data on which it is supposedly based. According to Coughley (1970), eruptive fluctuations of ungulates – an increase in numbers over at least two generations, followed by a marked decline – are initiated by a change in food or habitat, and terminated by overgrazing. The Kaibab eruptions fits this pattern. The increase in numbers was certainly correlated with the removal of predators; but competitors – sheep and cattle – were also removed. Not only had grazing altered the habitat, but fire had also played a part. Reduction in numbers of predators may have had minor influence.

No area can support a population above its 'carrying capacity', a term that has been used in at least three senses, but which has been defined by Dasmann (1945) as the number of animals that a habitat can maintain in a healthy vigorous condition. It is very important to determine the carrying capacity of land in order that the wild life may be conserved with maximum efficiency. The principles of wildlife biology and game conservation are outlined by Dasmann (1964a; 1964b), Leopold (1933) and others. They lie outside the scope of the present book. We will, therefore, conclude this chapter by reference to a number of cases in which changes in the numbers of herbivorous species have influenced considerably the vegetation and soils of their habitats. Some of the more harmful of these influences might well have been avoided by the application of appropriate methods of conservation.

Of the great forests that once covered most of the British Isles, only fragments remain and most of these have been much modified by man. As Tansley (1939) points out, most of our remaining natural and semi-natural communities are dominated by oak, beech, ash, birch, pine or alder. They tend to be sharply limited and enclosed by agricultural land, in consequence of which their natural spread is prevented and they are vulnerable to invasion by 'weeds' wherever open, well lighted soil occurs. Regeneration is therefore inhibited. Secondly, there is a great diminution in the number of parent plants to produce seed; and destruction of carnivorous mammals and birds by game-keepers has led to an enormous increase in the numbers of small herbivorous voles, mice, squirrels and rabbits which, in turn, destroy seeds and seedling trees to such an extent that the capacity of the natural woodland to maintain itself is seriously impaired.

Rabbits. Until the advent of myxomatosis, grazing by rabbits *(Oryctolagus cuniculus)* was probably the most widespread and effective biotic factor in modifying Britain's semi-natural vegetation. The rabbit, originally a native of south-western Europe, was introduced into England, probably by the Normans in the twelfth century. At first they were scarce but, by the middle of the fifteenth century they had become abundant and were highly valued as a source of food and fur.

Although they do not range very far from their burrows, the effect of rabbits on the vegetation can be very marked. Especially in chalkland, the soil may be quite bare immediately around their burrows. A little distance off, the plants of the turf are eaten down to a height of about a centimetre over a considerable area. Farther away, the effect of grazing is less severe but it may be noticeable up to 150 m. In many areas, burrows were so close together that severely grazed areas overlapped.

Rabbits are selective in their choice of food plants and will not touch bracken fern *(Pteridium aquilinum)*, heaths *(Erica* spp.*)* and various other species which flourish in heavily grazed areas.

In breckland, the thinning or destruction of the vegetation by rabbits may give the wind a purchase on the underlying sand which is blown away. The same effect is enhanced in coastal dunes which may become severely eroded (Tansley, 1939). After reduction of rabbit populations as a result of the introduction of myxomatosis into Britain, and their subsequent stabilisation at a much lower level than heretofore, there has been considerable regeneration of woody scrub on the downland sward of this country.

Caribou. The relationships of wildlife to biotic succession are most striking in forested areas because here, as Dasmann (1964b) points out, the greatest difference exists between the climax community and the open, pioneer communities of bare ground. A striking example of these relationships is provided in the northern forests of the New World which form the winter range of the barren-ground caribou *(Rangifer tarandus)*. These animals feed on a mixture of climax of lichens *(Cetraria* and *Cladonia* spp.*)* which are associated with undisturbed taiga. The lichens grow very slowly because precipitation is low and the growing season short; but caribou do not exert heavy grazing pressure since they are constantly on the move. When domesticated reindeer were introduced, however, serious damage was caused because their Eskimo herders failed to keep the animals on the move. Large areas of former caribou range in Alaska were thus put out of production through destruction of the lichens (Lepold and Darling, 1953). (p. 138).

Hippopotamus. Before the advent of fire-arms, the game herds of tropical Africa had their numbers regulated by predators, including man and, in some cases, by shortage of water holes. When the numbers of carnivores were reduced by hunters, however, there was a consequent increase in the populations of herbivores in certain areas. Thus, huge populations of *Hippopotamus amphibius* built up in the Victoria Nile, and around Lake George and Lake Edward, where they were protected. Their nocturnal grazing habits, coupled with the considerable amount of faecal waste matter discharged into the water during the day, results in a continual one-way drain of nutrients from the terrestrial ecosystem. This has an adverse effect on the habitat in areas of high grazing density and causes gully erosion (Laws and Clough, 1965). When hippopotamuses are removed by shooting, the vegetation changes and various species of grass become re-established (Thornton, 1971). By their movements to and fro, hippopotamuses in West Africa may keep streams free from excessive growths of weed and thus reduce flooding of the land (Clarke, 1953).

Elephant. One of the most important problems of wildlife management in Africa concerns the elephant *(Loxodonta africana).* While the total population is insignificant compared with the vast numbers which must have existed in previous centures, numbers have recently increased very considerably in certain localities, notably the Murchison Falls and Tsavo National Parks, where elephants have greatly reduced the vegetation cover. According to Sikes (1966; 1971), the Murchison elephants have increased in an area of abundant natural surface water and considerable amounts of vegetation, but here they have to remain because of human encroachment on their natural migration routes and alternative refuges. Within the area, the combined effects of elephants and fire have greatly reduced the vegetation cover, but the situation is less serious than at Tsavo where it has been aggravated by the introduction of permanent water holes by the Parks authorities. Overcrowding is particularly severe in times of drought (Glover, 1963). The seriousness of the problem has been stressed by Laws (1970) who presented a number of case histories of elephant populations and habitat change. He discussed the nature of the behavioural changes associated with contraction of range and damage to the habitat.

Laws, Parker and Johnstone (1970) analysed the populations of elephants in the Murchison Falls National Park and surrounding areas and concluded that they are at densities which exceed the carrying capacity of the habitats. In areas of grassland and savannah woodland, the destruction of trees has progressed radially, a zone of damage about 15 – 20 km wide having moved outwards through the range. This is consistent with destruction mainly by elephants rather than by fire. Although self-regulatory mechanisms – reduced fertility and increased mortality of the calves – come into operation, they would appear to be inadequate to prevent a crash in the population in the absence of a third regulatory mechanism, namely emigration (Laws and Parker, 1968). There would appear to be little justification for the complacency expressed by Glover (1970).

Even though the impact of elephants on the vegetation is less dramatic in Asia than in Africa, it is interesting to record that Indian elephants *(Elephas maximus)* are responsible for distortion of the crowns of woody plants in the Ruhuna National Park of Sri Lanka

(Ceylon), according to Mueller-Dombois (1971). If the population were to build up through conversion of the natural habitat into agricultural development schemes, the still well-established forest could readily break down from over-use.

In this chapter I have given only a few examples of the ways in which animals can influence the vegetation and, through it, the soil and possibly even the climate. I have cited some of the more spectacular cases involving mammals, but have not mentioned insects which can entirely defoliate trees such as oaks (Elton, 1966) or transmit pathogenic organisms.

15 ECOLOGICAL REGULATION

All living organisms require nutrients, space, and the opportunity to reproduce themselves. These demands must be met by the environment which provides radiant energy (fixed by green plants in photosynthesis), inorganic salts, carbon dioxide and water. The properties of the environment that permit organisms to survive are not distributed evenly, however, and this results in the non-uniform distribution of living things. Resources also vary greatly in quality as well as in quantity, and there are, therefore, great evolutionary pressures towards specialisation in requirements. These pressures derive from competition between individuals, predation, and from the heterogeneous nature of the environment: they result in the profusion of species seen on earth today (Healey, 1972).

Within the various ecosystems of the world, vegetation, micro-organisms, fauna, climate and soil interact with one another through countless feed-back systems. Olson (1972) points out that, in its evolution, life has not only responded to, but also has probably caused, major climatic changes. Green plants have polluted the primeval atmosphere with oxygen, as R.N. T-W-Fiennes explains in the first volume of this series, thereby making the earth a suitable environment for habitation by animals which, with plants, play vital roles in the cycles of nitrogen, carbon, sulphur and other essential elements. Living organisms are also, to a large degree, responsible for the production of the humus and soil that render the earth suitable for other plants and animals. These points are well illustrated by Sukacher and Dylis (1968) and by Reichle (1970) in the case of forests, and by Spedding (1971) in that of grasslands; while the pattern of animal communities forms the subject of Elton's (1966) masterpiece.

The inter-relationships between climate, physical environment and living organisms are exceedingly diverse and complicated. For instance, the plants of Palaearctic chalk downs and limestone crags are a mixture. Some could be seen equally well in other places; others, and this applies particularly to the rarities, are calcicoles and are either restricted to, or grow specially well on, soils containing a high percentage of lime (Lousley, 1950). Many of the herbs and shrubs of calcareous hills in Pakistan also have a restricted distribution and are likewise characteristic of particular localities: they are very resistant to drought (Shaukat and Quadir, 1971). In addition, northerly slopes, which are cooler than southern aspects, have a denser vegetation. Thus, parent rock and direction of slope are both factors influencing the flora and, secondarily, the fauna.

The interdependence of plants and animals has been discussed by Gunderson and Hastings (1944) who pointed out that plants have become adjusted to the activities of herbivores by various regenerative and protective devices. Because they grow from the base of the leaf, grasses are able to withstand grazing and thus dominate steppe whence other types of plants, which grow at the end of the stem, are largely eliminated. Cacti and other desert succulents survive in arid areas, where the plant population is reduced by shortage of water, because their spines give them protection from the larger herbivores. Cattle readily eat prickly pear *(Opuntia inermis)* when the spines have been burned off, and camels will eat it spines and all (Allee *et al.*, 1949). Perhaps this is not coincidence since Cactaceae originated in the New World while modern camels are Palaearctic. Other desert plants have evolved repellant taste, smell or purgative effects that can be as protective as spines!

As I mentioned in Chapter 3, the numbers of invertebrates inhabiting tropical rain-forest are low, and the diversity of species is very great, so the population density of most species must be extremely low. Elton (1973) has recently propounded a theory that this low population density may be the end-product of a very long interaction between predators and their prey. As a result of this, the prey species that survive also live mostly at very low population density or are nocturnal. This historical process is thought to have been assisted by local group extinctions occurring in species already living at low population levels. A rather fragile system of this kind may well need a very large area of rain-forest for its long-term survival. Defensive and offensive adaptations have often reached perfection in tropical forest. An animal cannot, however, look more like a leaf or twig than one twig or leaf looks like another. 'The same goes for mimicry of other animals. Even chemical and other defences, perhaps accompanied by aposematic characteristics, are not effective against all predators. Suppose, then, we imagine a great many defence codes made and then the codes broken, until those lines of evolutionary adaptation simply cannot be developed any further. There would seem to be three possible further directions to take: (a) extinction; (b) change to nocturnal habits (cf. Elton, 1966, p.25); (c) to live at very low population density.' Elton (1973) argues that all three have taken place.

Animals depend upon plants directly or indirectly for carbohydrates, certain essential amino-acids and certain vitamins. In their turn, plants are dependent, to a degree, upon the activities of animals through the parts they play in the nitrogen, carbon and phosphorus cycles, through their geological influence in modifying the soil and their roles as seed dispersal and pollinating agents. A balanced equilibrium between plants and animals is favourable to both. In this book, however, I have not dealt with energy flow, populations and certain others aspects of ecology not directly related to the environment. Nor have I discussed, in detail, the ways in which man has upset the balance (Ehrlich and Ehrlich, 1970).

A considerable amount of general evidence suggests that there is an optimum productivity for a particular environment. The rate of growth of plants depends not only on various factors, such as soil fertility, competition and exposure, but is adpated to particular environments. For example, Arctic-Alpine plants are usually species of low productivity, while those of rich grasslands are highly productive. Bradshaw (1957) has demonstrated that, in some of the man-made grasslands of England, which have been overgrazed for centures, *Agrostis tenuis* develops ecotypes with inherent low productivity even when transplanted to ungrazed experimental gardens. Presumably it is of survival value for plants to grow slowly where they are heavily grazed, even when moisture, light and temperature are favourable. In this way they avoid being killed completely by grazing animals.

Terrestrial habitats are almost infinitely variable, and the plant and animal communities dependent on them also vary greatly in their mode of life. Apart from physical modifications, there are an infinite number of physiological adaptations that lie beyond the scope of this book. Where plants grow well on soils that are deficient in elements essential to the well-being of animals, such as calcium, phosphorus, iron, manganese, molybdenum, cobalt and so on, the fauna is naturally poor. The nitrogen sources for all living material are derived largely from the atmosphere and made available by nitrogen-fixing bacteria in the soil and the roots of leguminous plants. This is yet another way in which animals depend on plants which, in turn, rely in more ways than this on soil-living organisms (Fiennes, 1972).

The machinery by which homeostasis is achieved includes the co-ordinated physiological processes that maintain most of the steady states in organisms. These have scarcely been mentioned in the present volume. Similar general principles, however, apply to the establishment, regulation and control of steady states for other levels of organisation (Pantin, 1964). As Henderson (1913) pointed out, the physical properties of the environment present a unique collection of properties essential for the maintenance of that class of systems to which living organisms belong. These properties cover different qualities, from the special properties of carbon and of water to the special set of physical conditions on this planet. There is no obvious linkage between many of these special conditions: that is, given that one is true, we have no reason to suppose that the truth of the others could be inferred from it.

'This does not imply that systems with certain essential features characteristic of life as we know it may not exist in different conditions elsewhere in the universe – as, for instance, conceivably in liquid ammonia systems in the outer planets – or even in certain classes of machine to be constructed on our own – but that the possibility of the whole set of these depends upon unique properties of the material universe which show no necessary linkage' (Pantin, 1964).

Since Henderson's day, the number of known unique properties favourable for the existence of living organisms has increased, from the remarkable properties of the DNA molecule to those of sodium by which the cellular ionic composition is maintained. In discussing natural selection, Darwin (1859) wrote: 'Let it be borne in mind how infinitely close-fitting are the mutual relations of all organic beings to each other and to their physical conditions of life.' Pantin (1968) emphasised that these 'close-fitting' relations are the result of natural selection and that, while biologists have given much thought to the adaptations of the living organism to the environment, both before and after the publication of Henderson's (1913) book they have paid little attention to the nature of the environment itself. In pre-Darwinian days, however, this was not so, and the peculiar fitness of the environment for living things was as well recognised as was the apparent element of design in organisms themselves.

In this book, attention has been focussed on virgin ecosystems of the terrestrial environment in which man plays little or no part. Plants, animals and environment have evolved together and are interdependent in approximately stable and homeostatic equilibrium. It is this stability that is so often disturbed by human influences which may thereby imperil life itself.

BIBLIOGRAPHY

This bibliography is by no means comprehensive, but has been limited to works actually cited in the text. Nevertheless, I hope that it will provide a useful and representative entree to the literature on the various aspects of ecology discussed in the preceeding pages. References to non-zoological papers have been restricted as far as possible, to books and reviews concerned with the whole world, rather than to research papers or regional studies. Zoological publications inevitably reflect my own interests: for this reason, too, my own work has been quoted more frequently than could possibly be justified otherwise.

Ahearn, G.A. (1970), 'The control of water loss in desert tenebrionid beetles', *J. Exp. Biol.,* **53**, pp. 573-95.

Allee, W.C., Emerson, A.E., Park, O., Park, T. and Schmidt, K.P. (1949), *Principles of Animal Ecology,* Saunders, Philadelphia, 837 pp.

Allen, G.M. (1939), *Bats,* Harvard Univ. Press, Cambridge, Mass., 368 pp.

Alluaud, C. (1908), 'Les coleopteres de la faune alpine du Kilimandjaro avec notes sur la faune du Mont Meru', *Ann. Soc. ent. France,* **77**, pp. 21-32.

Andrewartha, H.G. (1952), 'Diapause in relation to the ecology of insects', *Biol. Rev.,* **27**, pp. 50-107.

Aslyng, H.C. (1958), 'Shelter and its effect on climate and water balance', *Oikos.* **9**, pp. 282-310.

Ahearn, G.A. and Hadley, N.F. (1969), 'The effects of temperature and humidity on water loss in two desert tenebrionid beetles, *Eleodes armata* and *Cryptoglossa verricosa', Comp. Biochem.Physiol.,* **30**, pp. 739-49.

Bagnold, R.A. (1954), 'The physical aspects of dry deserts'. *In* Cloudsley-Thompson, J.L. (ed.), *Biology of Deserts,* Inst. Biol., London, pp. 7-12.

Baker, J.R. (1938), 'The evolution of breeding seasons'. *In* de Beer, G.R. (ed.), *Evolution,* Oxford Univ. Press, London, pp. 161-77.

Baker, J.R. and Baker, Z. (1936), 'The seasons in a tropical rain-forest (New Hebrides). Part 3. Fruit-bats (Pteropidae)', *J. Linn, Soc. (Zool.),* **39**, pp. 123-41.

Baker, J.R. and Bird, T.F. (1936), 'The seasons in a tropical rain-forest (New Hebrides), Part 4. Insectivorous bats (Vespertilionidae and Rhinolophidae)', *J. Linn. Soc. (Zool.),* **40**, pp. 143-61.

Baker, R.R. (1968), 'A possible method of evolution of the migratory

habit in butterflies', *Phil. Trans. Roy. Soc.*, (B) **253**, pp. 309-41
Baker, R.R. (1969), 'The evolution of the migratory habit in butterflies', *J. Anim. Ecol.*, **38**, pp. 703-46.
Barnes, H.F. (1944), 'Discussion on slugs. 1. Introduction. Seasonal activity of slugs', *Ann. Appl. Biol.*, **31**, pp. 160-3.
Bartholomew, G.A. and Dawson, W.R. (1956), 'Temperature regulation in the macropod marsupial, *Setonix brachyurus*', *Physiol. Zoöl*, 29, pp. 26-40.
Bates, H.W. (1863), *The Naturalist on the River Amazons,* Murray, London, 466 pp.
Beadle, L.C. (1963), 'Anaerobic life in a tropical crater lake', *Nature, Lond.*, 200, pp. 1223-4.
Beard, J.S. (1953), 'The savanna vegetation of northern tropical America', *Ecol. Monogr.*, **23**, pp. 149-315.
Beaufort, L.F. de (1951), *Zoogeography of the Land and Inland Waters,* Sidgewick and Jackson, London, 208 pp.
Beck, S.D. (1968), *Insect Photoperiodism,* Academic Press, New York, 288 pp.
Beebe, C.W., Hartley, G.I. and Hawes, P.G. (1917), *Tropical Wild Life in British Guiana,* New York Zool. Soc., New York, 504 pp.
Belcher, C.F. (1930), *The Birds of Nyasaland,* Crosbey Lockwood, London, 356 pp.
Belt, T. (1874), *The Naturalist in Nicaragua,* Murray, London, 403 pp.
Benson, S.B. (1933), 'Concealing coloration among some desert rodents of the South-western United States', *Univ. Calif. Publ. Zool.*, **40**, pp. 1-70.
Bernard, F. (1948), 'Les Insectes sociaux du Fezzan. Comportement et Biogeographie'. *In Mission Scientifique du Fezzan (1944-1945) V Zoologie (Arthropodes, I),* pp. 85-201, Inst. Recherches sahariennes de l'Univ. d'Alger, Alger.
Bert, W.H. and Grossenheider, R.P. (1952), *A Field Guide to the Mammals.* Houghton Mifflin, Boston 200 pp.
Billard, G. and Bruyant, G. (1905), 'Sur un mode particulier de locomotion de certains *Stenus*', *C.R. Soc. Biol.*, no. **59**, pp. 102-3.
Bodenheimer, F.S. (1957), 'The ecology of mammals in arid zones'. *In Human and Animal Ecology, UNESCO,* Paris, pp. 100-137.
Bird, R.D. (1930), 'Biotic communities of the aspen parklands of central Canada', *Ecology,* **11**, pp. 356-442.
Black, C.A. (1957), *Soil-plant Relationships,* Wiley, New York, 332 pp.
Blest, A.D. (1966), 'Evolutionary relationships between insects and their predators'. *In* Kalmus, H. (ed.), *Regulation and Control in Living Systems,* Wiley, London, pp. 380-96.
Bodenheimer, F.S. (1954), 'Problems of physiology and ecology of desert animals'. *In* Cloudsley-Thompson, J.L. (ed.), *Biology of*

Deserts', Inst. Biol., London, pp. 162-7.
Bouliere, F. and Hadley, M. (1970), 'The ecology of tropical savannas', *Ann. Rev. Ecol. Syst.*, **1**, pp. 125-52.
Bradshaw, A.D. (1957), 'The genecology of productivity of grasses' (Abstract), *J. Anim. Ecol.*, **26**, pp. 242.
Brinck, P. (1956), 'The food factor in animal desert life'. *In* W.G. Wingstrand (ed.), *Bertil Hanstrom: Zoological Papers in Honour of his 65th Birthday, November 20th 1956*, Zool. Inst., Lund, pp. 120-37.
Brooks, A.C. (1961), 'A study of the Thomson's gazelle *(Gazella thomsonii* Gunther) in Tanganyika', *Colonial Res. Publ.* no. 25, pp. 1-147.
Brower, L.P. and Brower, J.V. (1972), 'Parallelism, convergence, divergence, and the new concept of advergence in the evolution of mimicry', *Trans. Conn. Acad. Arts Sci.*, **44**, pp. 59-67.
Brower, L.P., Cook, L.M. and Croze, H.J. (1967), 'Predator responses to artificial Batesian mimics released in a Neotropical environment', *Evolution*, **21**, pp. 11-23.
Bullough, W.S. (1952), *Vertebrate Sexual Cycles*, Methuen, London, 117 pp.
Bump, G. Darrow, R.W., Edminster, F.C. and Crissey, W.F. (1947), *The Ruffed Grouse: Life History, Propagation, Management*, State Cons. Dept., New York, 915 pp.
Buxton, P.A. (1923), *Animal Life in Deserts*, Arnold, London, 176 pp.
Caborn, J.M. (1973), 'Microclimates', *Endeavour*, **32**, pp. 30-33.
Cade, T.J. (1953), 'Sub-nival feeding of the redpoll in interior Alaska: a possible adaptation to the northern winter', *Condor*, **55**, pp. 43-4.
Cade, T.J. and Maclean, G.L. (1967), 'Transport of water by adult sandgrouse to their young', *Condor*, **69**, pp. 323-43.
Calvert, A.S. and Calvert, P.P. (1917), *A Year of Costa Rican Natural History*, Macmillan, New York, 577 pp.
Carlisle, D.B., Ellis, P.E. and Betts, E. (1965), 'The influence of aromatic shrubs on sexual maturation in the desert locust *Schistocerca gregaria*', *J. Insect Physiol.*, **11**, pp. 1541-58.
Carlquist, S. (1966), 'The biota of long-distance dispersal. 1. Principles of dispersal and evolution', *Quart. Rev. Biol.*, **41**, pp. 247-70.
Carpenter, K.E. (1928), *Life in Inland Waters with Especial Reference to Animals*, Sidgwick and Jackson, London, 267 pp.
Caughley, G. (1970), Eruption of ungulate populations, with emphasis on Himalayan Thar in New Zealand, *Ecology*, **51**, pp. 53-72.
Caullery, M. (1952), *Parasitism and Symbiosis* (Trans. A.M. Lysaght), Sidgwick and Jackson, London, 340 pp.
Chapin, J.P. (1932), 'The birds of the Belgian Congo, Part 1', *Bull. Amer. Mus. Nat. Hist.*, **65**, pp. 1-756.

Chapman, B.M. and Chapman, R.F. (1958), 'A field study of a population of leopard toads *(Bufo regularis regularis)', J. Anim. Ecol.,* **27**, pp. 265-86.

Chapman, B.M., Chapman, R.F. and Robertson, I.A.D. (1959), 'The growth and breeding of the multimammate rat, *Rattus (Mastomys) natalensis* (Smith) in Tanganyika Territory', *Proc. Zool. Soc. Lond.,* **133**, pp. 1-9.

Chapman, R.N. (1931), *Animal Ecology with Especial Reference to Insects,* McGraw-Hill, New York, 464pp.

Chapman, R.N. Mickel, C.E. Parker, J.R. *et al.* (1926), 'Studies in the ecology of sand dune insects', *Ecology,* **7**, pp. 416-26.

Chitty, H. (1950) 'Canadian Arctic wild life enquiry, 1943-49: with a summary of results since 1933', *J. Anim. Ecol.,* **19**, pp. 180-93.

Clarke, J.R. (1953), 'The hippopotamus in Gambia, West Africa', *J. Mammal,* **34**, pp. 299-315.

Clayton, W.D. (1963), 'The vegetation of Katsina Province, Nigeria', *J. Ecol.,* **51**, pp. 345-51.

Clements, F.E. and Shelford, V.E. (1939), *Bio-ecology,* Wiley, New York, 425 pp.

Cloudsley-Thompson, J.L. (1952), 'Studies in diurnal rhythms. II Changes in the physiological responses of the woodlouse *Oniscus asellus* to environmental stimuli', *J. Exp. Biol.,* **29**, pp. 295-303.

Cloudsley-Thompson, J.L. (1956a), 'Studies in diurnal rhythms. VI. Bioclimatic observations in Tunisia and their significance in relation to the physiology of the fauna, especially woodlice, centipedes, scorpions and beetles', *Ann. Mag. Nat. Hist.,* (12) **9**, pp. 305-29.

Cloudsley-Thompson, J.L. (1956b), 'Studies in diurnal rhythms. VII. Humidity responses and nocturnal activity in woodlice (Isopoda)', *J. Exp. Biol.,* 33, pp. 276-82.

Cloudsley-Thompson, J.L. (1959), 'Studies in diurnal rhythms. IX. The water relations of some nocturnal tropical arthropods', *Ent. Exp. & Appl.,* **2**, pp. 249-56.

Cloudsley-Thompson, J.L. (1960a), *Animal Behaviour,* Oliver and Boyd, Edinburgh, 162 pp.

Cloudsley-Thompson, J.L. (1960b), 'Adaptive functions of circadian rhythms', *Cold Spr. Harb. Symp. Quant. Biol.,* **25**, pp. 345-55.

Cloudsley-Thompson, J.L. (1961a), *Rhythmic Activity in Animal Physiology and Behaviour,* Academic Press, New York and London, 236 pp.

Cloudsley-Thompson, J.L. (1961b), 'Observations on the natural history of the 'camel-spider', *Galeodes arabs* C.L.K. (Solifugae: Galeodidae) in the Sudan', *Entomologist's Mon. Mag.,* **97**, pp. 45-52.

Cloudsley-Thompson, J.L. (1961c), 'Some aspects of the physiology and behaviour of *Galeodes arabs', Ent. Exp. & Appl.,* **4**, pp. 257-63.

Cloudsley-Thompson, J.L. (1962a), 'Microclimates and the distribution of terrestrial arthropods', *Ann. Rev. Ent.*, **7**, pp. 199-222.
Cloudsley-Thompson, J.L. (1962b), 'Some aspects of the physiology and behaviour of *Dinothrombium* (Acari)', *Ent. Exp. & Appl.*, **5**, pp. 69-73.
Cloudsley-Thompson, J.L. (1962c), 'Bioclimatic observations in the Red Sea hills and coastal plain, a major habitat of the desert locust', *Proc. R. Ent. Soc. Lond.*, (A) **37**, pp. 27-34.
Cloudsley-Thompson, J.L. (1962d), 'Lethal temperatures of some desert arthropods and the mechanism of heat death', *Ent. Exp. & Appl.*, **5**, pp. 270-80.
Cloudsley-Thompson, J.L. (1963a), 'A note on the association between *Bengalia* spp. (Dipt., Calliphoridae) and ants in the Sudan', *Entomologist's Mon. Mag.*, **98**, pp. 177-9.
Cloudsley-Thompson, J.L. (1963b), 'Light responses and diurnal rhythms in desert Tenebrionidae', *Ent. Exp. & Appl.*, **6**, pp. 75-8.
Cloudsley-Thompson, J.L. (1964a), 'Terrestrial animals in dry heat: arthropods'. *In* Dill, D.B. (ed.), *Handbook of Physiology*, Sect. 4, pp. 451-65, Amer. Physiol. Soc., Washington, D.C.
Cloudsley-Thompson, J.L. (1964b), 'On the function of the sub-elytral cavity in desert Tenebrionidae', *Entomologist's Mon. Mag.*, **100**, pp. 148-51.
Cloudsley-Thompson, J.L. (1965a), *Desert Life*, Pergamon, Oxford, 86 pp.
Cloudsley-Thompson, J.L. (1965b), 'The scorpion', *Sci. J.*, **1** (5), pp. 35-41.
Cloudsley-Thompson, J.L. (1965c), 'Rhythmic activity, temperature-tolerance, water-relations and mechanism of heat death in a tropical skink and gecko', *J. Zool., Lond.*, **146**, pp. 55-69.
Cloudsley-Thompson, J.L. (1965d), *Animal Conflict and Adaptation*, Foulis, London, 160 pp.
Cloudsley-Thompson, J.L. (1966), 'Seasonal changes in the daily rhythms of animals', *Inst. J. Biometeor.*, **10**, pp. 119-25.
Cloudsley-Thompson, J.L. (1967a), *Microecology*, Arnold, London, 49 pp.
Cloudsley-Thompson, J.L. (1967b), *Animal Twilight. Man and Game in Eastern Africa*, Foulis, London, 204 pp.
Cloudsley-Thompson, J.L. (1967c), 'Diurnal rhythm, temperature and water relations of the African toad. *Bufo regularis*', *J. Zool., Lond.*, **152**, pp. 43-54.
Cloudsley-Thompson, J.L. (1968), *Spiders, Scorpions, Centipedes and Mites* (2nd ed.), Pergamon, Oxford, 278 pp.
Cloudsley-Thompson, J.L. (1969), *The Zoology of Tropical Africa*, Weidenfeld & Nicolson, London, 355 pp.

Cloudsley-Thompson, J.L. (1970a), 'Recent work on the adaptive functions of circadian and seasonal rhythms in animals', *J. Interdiscipl. Cycle Res.*, **1**, pp. 5-19.
Cloudsley-Thompson, J.L. (1970b), 'Terrestrial invertebrates'. *In* G.C. Whitton (ed.), *Comparative Physiology of Thermoregulation,* Academic Press, New York, **1**, pp. 15-77.
Cloudsley-Thompson, J.L. (1971a), *The Temperature and Water Relations of Reptiles,* Merrow, Watford, Herts, 159 pp.
Cloudsley-Thompson, J.L. (1971b), 'Non-adaptive variation and group selection', *Sci. Prog., Oxf.,* **59**, pp. 243-54.
Cloudsley-Thompson, J.L. (1971c), 'Recent expansion of the Sahara', *Intern J. Environmental Sci.,* **2**, pp. 35-9
Cloudsley-Thompson, J.L. (1972a), 'The habitat and its influence on the evolutionary development of life forms', *In* T.-W.-Fiennes, R.N. (ed.), *Biology of Nutrition,* Pergamon, Oxford, pp. 351-72.
Cloudsley-Thompson, J.L. (1972b), 'Temperature regulation in desert reptiles', *Symp. Zool. Soc. Lond.,* No. 31, pp. 39-59.
Cloudsley-Thompson, J.L. (1973), 'Factors influencing the supercooling of tropical Arthropoda, especially locusts', *J. Nat. Hist.,* **7**, pp. 481-80.
Cloudsley-Thompson, J.L. (1975a), *The Water and Temperature Relations of Woodlice* (Isopoda: Oniscoidea), Merrow, Watford, Herts.
Cloudsley-Thompson, J.L. (1975b), *The Ecology of Oases,* Merrow, Watford, Herts.
Cloudsley-Thompson, J.L.(1975c), *Evolutionary Trends in the Mating of Arthropoda,* Merrow, Watford, Herts.
Cloudsley-Thompson, J.L. (1975d), *The Size of Animals,* Merrow, Watford, Herts.
Cloudsley-Thompson, J.L. (1975e), *Aspects of Form and Function in Animals,* Merrow, Watford, Herts.
Cloudsley-Thompson, J.L. and Chadwick, M.J. (1964), *Life in Deserts,* Foulis, London, 218 pp.
Cloudsley-Thompson, J.L. and Mohamed, E.M. (1967), 'Water economy of the ostrich', *Nature, Lond.,* **212**, p. 306.
Coe, M.J. (1967), *The Ecology of the Alpine Zone of Mount Kenya,* Junk, The Hague, 136 pp.
Corbet, P.S. (1962), *A Biology of Dragonflies,* Witherby, London, 247 pp.
Corbet, P.S. (1967), 'Further observations on diel periodicities of weather factors near the ground at Hazen Camp, Ellesmere Island, N.W.T.', *Defence Research Board of Canada, Dept. Nat. Def. D. Phys. R. (G) Hazen,* **31**, 14 pp.
Corbet, P.S. (1972), 'The microclimate of arctic plants and animals, on

land and in fresh water', *Acta Arctica,* Fasc. 18, pp. 1-43.
Costello, D.F. (1964), 'Range dynamics control – an ecological urgency'. *In* Crisp, D.J. (ed.), *Grazing in Terrestrial and Marine Environments, Symp. Brit. Ecol. Soc.,* **4**, pp. 91-107.
Cott, H.B. (1930), 'The natural history of the lower Amazon', *Proc. Bristol Nat. Soc.,* (4) **7**, pp. 181-8.
Cott, H.B. (1940), *Adaptive Coloration in Animals,* Methuen, London, 508 pp.
Cott, H.B. (1961), 'Scientific results of an inquiry into the ecology and economic status of the Nile crocodile *(Crocodilus niloticus)* in Uganda and Northern Rhodesia', *Trans. Zool. Soc. London.,* **29**, pp. 211-356.
Cowles, R.B. (1957), 'Sidewinding locomotion in snakes', *Copeia,* **1956**, pp. 211-4.
Cowles, R.B. (1959), *Zulu Journal,* Univ. Calif. Press, Berkeley, 267 pp.
Cracraft, J. (1973), 'Continental drift, paleoclimatology, and the evolution and biogeography of birds', *J. Zool. Lond.,* **169**, pp. 455-545.
Curran, C.H. and Kauffeld, C. (1937), *Snakes and their Ways,* Harper, New York, 285 pp.
Cushing, D.H. (1951), 'The vertical migration of planktonic Crustacea', *Biol. Rev.,* **26**, pp. 158-92.
Dammermann, K.W. (1922), 'The fauna of Krakatau, Verlaten Island, and Sebesy', *Treubia,* 3, pp. 61-112.
Darlington, P.J. jr. (1957), *Zoogeography,* Wiley, New York, 675 pp.
Dasmann, R.F. (1964a), *African Game Ranching,* Pergamon, Oxford, 75 pp.
Dasmann, R.F. (1964b), *Wildlife Biology,* Wiley, New York, 231 pp.
Dasmann, W.P. (1945), 'A method for estimating carrying capacity of range lands', *J. Forestry,* **43**, pp. 400-2.
Darwin, C. (1859), *On the Origin of Species by Means of Natural Selection,* Murray, London, 502 pp.
Delany, M.J. (1964), 'An ecological study of the small mammals in the Queen Elizabeth Park, Uganda', *Rev. Zool. Bot. Afr.,* **70**, pp. 129-47.
Delany, M.J. (1971), 'The biology of small rodents in Mayanja Forest, Uganda', *J. Zool., Lond.,* **165**, pp. 85-129.
Delany, M.J. (1972), 'The ecology of small rodents in tropical Africa', *Mammal Rev.,* **2**, pp. 1-42.
Den Boer, P.J. (1961), 'The ecological significance of activity patterns in the woodlouse, *Porcellio scaber* Latr. (Isopoda)', *Archs. neerl. Zool.,* **14**, pp. 283-409.
Dierterlen, F. (1967), 'Jahreszeiten und Fortpflanzungsperioden bei den Muriden des Kivusee-Gebietes (Congo). Teil 1. Ein Beitrag zum Problem der Populationsdynamik in den Tropen', *Z. Saugetierk.,*

32, pp. 1-44.
Dietz, R.S. and Holden, J.C. (1970a), 'Reconstruction of Pangaea: breakup and dispersion of continents, Permian to present', *J. Geophys. Res.*, **75**, pp. 4939-56.
Dietz, R.S. and Holden, J.C. (1970b), 'The breakup of Pangaea', *Sci. Amer.*, **223** (4), pp. 30-41.
Dietz, R.S. and Holden, J.C. (1971), 'Pre-Mesozoic oceanic crust in the eastern Indian Ocean (Wharton Basin)', *Nature, Lond.*, **229**, pp. 309-12.
Dingle, H. (1972), 'Migration strategies of insects', *Science*, **175**, pp. 1327-35.
Dizer, Yu. B. (1955), 'On the physiological role of the elytra and sub-elytral cavity of steppe and desert Tenebrionidae', *Zool. zh. S.S.S.R.*, **34**, pp. 319-22 [In Russian].
Dorst, J.P. (1962), *The Migrations of Birds* (Trans. C.D. Sherman), Heinemann, London, 476 pp.
Downes, J.A. (1964), 'Arctic insects and their environment', *Can. Ent.*, **96**, pp. 279-307.
du Toit, A.L. (1937), *Our Wandering Continents*, Oliver & Boyd, Edinburgh, 366 pp.
Edney, E.B. (1965), 'Some aspects of adaptation in desert mammals', *Zoöl. Africana*, 1, pp. 1-8.
Edney, E.B. (1968), 'Transition from water to land in isopod crustaceans', *Amer. Zool.*, **8**, pp. 309-26.
Edney, E.B. (1971), 'The body temperature of tenebrionid beetles in the Namib Desert of southern Africa', *J. Exp. Biol.*, **55**, pp, 253-72.
Ehrlich, P.R. and Ehrlich, A.H. (1970), *Population Resources Environment*, Freeman, San Francisco, 383 pp.
Elgood, J.H., Sharland, R.E. and Ward, P. (1966), 'Palaearctic migrants in Nigeria', *Ibis*, **108**, pp. 84-116.
Elton, C. (1927), *Animal Ecology*, Sidgwick & Jackson, London, 207 pp.
Elton, C. (1930), *Animal Ecology and Evolution.* Clarendon Press, Oxford, 96 pp.
Elton, C.S. (1942), *Voles, Mice and Lemmings. Problems in Population Dynamics*, Clarendon Press, Oxford, 496 pp.
Elton, C.S. (1958), *The Ecology of Invasions by Animals and Plants*, Methuen, London, 181 pp.
Elton, C.S. (1966), *The Pattern of Animal Communities*, Methuen, London, 432 pp.
Elton, C.S. (1973), 'The structure of invertebrate populations inside neotropical rain forest', *J. Anim. Ecol.*, **42**, pp. 55-104.
Erkinaro, E. (1961), 'The seasonal change of the activity of *Microtus agrestis*', *Oikos*, **12**, pp. 157-63.
Estes, R.D. (1966), Behaviour and life history of the wildebeeste

(Connochaetes)', *Nature, Lond.*, **212**, pp. 999-1000.
Estes, R.D. (1967), 'Predators and scavengers', *Nat. Hist.*, **76**, pp. 21-9, 38-47.
Evans, A.H. (1899), *Birds,* Cambridge Natural History, Vol. IX, Macmillan, London, 635 pp.
Farner, D.S. and Follett, B.K. (1966), 'Light and other environmental factors affecting avian reproduction', *J. Anim. Sci.*, **25** (Suppl.), pp. 90-118.
Fiennes, R.N.T-W- (1972), 'The evolution and colonization of habitats'. *In* T-W-Fiennes, R.N. (ed.), *Biology of Nutrition,* Pergamon, Oxford, pp. 257-65.
Fisher, A.G. (1960), 'Longitudinal variations in organic density', *Evolution,* **14**, pp. 64-81.
Fitch, H.S. (1954), 'Life history and ecology of the five-lined skink, *Eumeces fasciatus'*, *Univ. Kansas Publ. Mus. Nat. Hist.*, **8**, pp. 1-156.
Fitch, H.S. (1970), 'Reproductive cycles of lizards and snakes', *Univ. Kansas Mus. Nat. Hist. Misc. Publ.*, No. 52, pp. 1-247
Folk, G.E. (1966), *Introduction to Environmental Physiology,* Lea & Febiger, Philadelphia, 308 pp.
Fraser, A.F. (1968), *Reproductive Behaviour in Ungulates,* Academic Press, London, 202 pp.
Frazer, J.F.D. (1959), *The Sexual Cycles of Vertebrates,* Hutchinson, London, 168 pp.
Friedlander, C.P. (1960), *Heathland Ecology,* Heinemann, London, 94 pp.
Frost, S.W. (1942), *General Entomology,* McGraw-Hill, New York, 524 pp.
Gadow, H. (1901), *Amphibia and Reptiles,* Cambridge Natural History, Vol. VIII, Macmillan, London, 668 pp.
Gadow, H. (1913), *The Wanderings of Animals,* Univ. Press, Cambridge, 150 pp.
Gartlan, J.S. and Struhsaker, T.T. (1972), 'Polyspecific associations and niche separation of rain-forest anthropoids in Cameroon, West Africa', *J. Zool., Lond.*, **168**, pp. 221-66.
Geiger, R. (1950), *The Climate near the Ground* (Trans. M.N. Stewart *et al.) (Das Klima der bodennahen Luftschicht),* Harvard Univ. Press. Cambridge, Mass., 482 pp.
George, W. (1962), *Animal Geography,* Heinemann, London, 142 pp.
Ghobrial, L.I. (1970), 'The water relations of the desert antelope *Gazella dorcas dorcas'*, *Physiol. Zoöl*, 43, pp. 249-56.
Gimingham, C.H. (1972), *Ecology of Heathlands,* Chapman & Hall, London, 266 pp.
Glover, J. (1963), 'The elephant problem at Tsavo', *E. Afr. Wildl. J.*, **1**, pp. 30-9.

Glover, P.E. (1970), 'The Tsavo and the elephants', *Oryx,* **10**, pp. 323-5.
Gough, D.I. Opdyke, N.D. and McElhinny (1964), 'The significance of paleomagnetic results from Africa', *J. Geophys. Res.,* **69**, pp. 2509-19.
Gray, J. (1968), *Animal Locomotion,* Weidenfeld & Nicolson, London, 479 pp.
Gunderson, A. and Hastings, G.T. (1944), 'Interdependence in plant and animal evolution', *Sci. Monthly,* **59**, pp. 63-72.
Haddow, A.J. (1952), 'Field and laboratory studies on an African monkey, *Cercopithecus ascenius* Matschie', *Proc. Zool. Soc. Lond.,* **122**, pp. 297-394.
Hadley, N.F. (1970), 'Micrometeorology and energy exchange in two desert arthropods', *Ecology,* **51**, pp. 434-44.
Hadley, N.F. (1972), 'Desert species and adaptation', *Amer. Sci.,* **60**, pp. 338-47.
Hafez, E.S.E. (1952), 'Studies on the breeding cycle and reproduction of the ewe', *J. Agric. Sci.* **42**, pp. 189-265.
Hafez, M. and Makky, A.M.M. (1959), 'Studies on desert insects in Egypt, III. On the bionomics of *Adesmia bicarinata* Klug.', *Bull, Soc. Ent. Egypte,* **43**, 89-113.
Hammerton, D. (1972), 'The Nile river – a case history' *In* Lee, D. and Oglesby, R. (ed.), *River Ecology and Man,* Academic Press, New York, pp. 171-214.
Hamilton, W.J. (1973), *Life's Color Code,* McGraw-Hill, New York 238 pp.
Hanney, P. (1964), 'The harsh-furred rat in Nyasaland', *J. Mammal.,* **45**, pp. 345-58.
Happold, D.C.D. (1966), 'Breeding periods of rodents in the northern Sudan', *Rev. Zool. Bot. Afr.,* **74**, pp. 357-63.
Happold, D.C.D. (1967), 'Biology of the jerboa, *Jaculus jaculus butleri* (Rodentia, Dipodidae), in the Sudan', *J. Zool., Lond.,* **151**, pp. 257-75.
Happold, D.C.D. (1968), 'Observations on *Gerbillus pyramidum* (Gerbillinae, Rodentia) at Khartoum, Sudan', *Mammalia,* **32**, pp. 44-53.
Harris, J.E. and Wolfe, U.K. (1955), 'A laboratory study of vertical migration', *Proc. Roy. Soc.,* (B) **144**, pp. 280-90.
Harrison, H. (1936), 'The Shinyanga game experiment: a few of the early observations', *J. Anim. Ecol.,* **5**, pp. 271-93.
Harrison, J.L. (1955), 'Data on the reproduction of some Malayan mammals', *Proc. Zool. Soc. Lond.,* **125**, pp. 445-60.
Harrison, J.L. (1957), 'Malaysian parasites – XXXIII. The hosts', *Stud. Inst. Med. Res. Malaya,* **28**, pp. 409-26.
Harrison, J.L. (1962), 'The distribution of feeding habits among animals in a tropical rain forest', *J. Anim. Ecol.,* **31**, pp. 53-64.

Haviland, M.D. (1926), *Forest, Steppe and Tundra,* Univ. Press, Cambridge, 218 pp.
Healey, I.N. (1972), 'The habitat, the community and the niche'. *In* T-W-Fiennes, R.N. (ed.), *Biology of Nutrition,* Pergamon, Oxford, pp. 307-49.
Heape, W. (1931), *Emigration, Migration and Nomadism,* Heffer, Cambridge, 369 pp.
Henderson, L.J. (1913), *The Fitness of the Environment,* Macmillan, New York, 317 pp.
Hershkovitz, P. (1969), 'The evolution of mammals on southern continents. VI. The recent mammals of the neotropical region: a zoogeographic and ecological review', *Quart. Rev. Biol.,* **44**, pp. 1-70.
Hesse, R. Allee, W.C. and Schmidt, K.P. (1951), *Ecological Animal Geography* (2nd ed.), Wiley, New York, 715 pp.
Hinton, H.E. (1951), 'A new Chironomid from Africa, the larva of which can be dehydrated without injury', *Proc. Zool. Soc. Lond.,* **121**, pp. 371-80.
Hinton, H.E. (1955), 'Protective devices of endopterygote pupae', *Trans. Soc. Brit. Ent.,* **12**, pp. 49-92.
Hinton, H.E. (1960), 'Cryptobiosis in the larva of *Polypedilum vanderplanki* Hint. (Chironomidae)', *J. Insect Physiol.,* **5**, pp. 286-300.
Hinton, H.E. (1968), 'Reversible suspension of metabolism and the origin of life', *Proc. Roy. Soc.* (B) **171**, pp. 43-56.
Hinton, H.E. (1971), 'Reversible suspension of metabolism', *C.R. 2nd Conf. Int. Phys. Théor. Biol.,* (1969), pp. 70-89.
Hock, R.J. (1964a), 'Terrestrial animals in cold: reptiles'. *In* Dill, D.B. (ed.), *Handbook of Physiology,* Sect. 4, pp. 357-9. Amer. Physiol. Soc., Washington, D.C.
Hock, R.J. (1964b), 'Animals at high altitudes: reptiles and amphibians', *In* Dill, D.B. (ed.), *Handbook of Physiology,* Sect. 4, 841-2, Amer. Physiol. Soc., Washington, D.C.
Hoffman, R.A. (1964), 'Terrestrial animals in cold: hibernators'. *In* Dill, D.B. (ed.), *Handbook of Physiology,* Sect. 4, pp. 379-403, Amer. Physiol. Soc., Washington, D.C.
Holdridge, L.R. (1947), 'Determination of world plant formations from sample climatic data', *Science,* **105**, pp. 367-8.
Hudson, J.W. and Bartholomew, G.A. (1964), 'Terrestrial animals in dry heat: estivators'. *In* Dill, D.B. (ed.), *Handbook of Physiology,* Sect. 4, pp. 379-403, Amer. Physiol. Soc., Washington, D.C.
Hutchinson, G.E. (1957), *A Treatise on Limnology,* Vol. I, *Geography, Physics and Chemistry,* Wiley, New York, 1015 pp.
Hutchinson, G.E. (1967), *A Treatise on Limnology,* Vol. II, *Introduction to Lake Biology and the Limnoplankton,* Wiley, New York, 1115 pp.

Ihering, H. von (1907), 'Die Cecropien und ihre Schutzameisen', *Engler's Botan. Jahrb.*, **39**, pp. 666-714.
Immelmann, K. (1963), 'Drought adaptations in Australian desert birds', *Int. Orn. Congr.*, **13**, pp. 649-57.
Imms, A.D. (1947), *Insect Natural History,* Collins, London, 317 pp.
Ionides, C.J.P. (1965), *A Hunter's Story,* Allen, London, 222 pp.
Irving, E. and Green, R. (1957), 'Paleomagnetic evidence from the Cretaceous and Cainozoic', *Nature, Lond.*, **179**, p. 1064.
Irving, E. Robertson, W.A. and Stott, P.M. (1963), 'The significance of the paleomagnetic results from Mesozoic rocks of Eastern Australia', *J. Geophys. Res.*, **68**, pp. 2313-7.
Iverson, S.L. Seabloom, R.W. and Hnatiuk, J.M. (1967), 'Small-mammal distributions across the prairie-forest transition of Minnesota and North Dakota', *Amer. Midl. Nat.*, **78**. pp. 188-97.
Jaeger, E.C. (1948), 'Does poor-will "hibernate"?', *Condor,* **50**, p.45.
Jaeger, E.C. (1965), *The California Deserts,* Stanford Univ. Press, Stanford, Calif., 208 pp.
Janzen, D.H. (1967), Interaction of the bull's horn acacia *(Acacia cornigera* L.*)* with an ant inhabitant *(Pseudomyrmex ferruginea* F. Smith*)* in eastern Mexico. *Univ. Kansas Sci. Bull,* **47**, pp. 315-558.
Janzen, D.H. (1973), Dissolution of mutalism between *Cecropia* and its *Azteca* ants. *Biotropica,* **5**, pp. 15-28.
Jarvis, J.U.M. (1969), 'The breeding season and litter size of African mole-rats', *J. Reprod. Fert. Suppl.*, No. 6, pp. 237-48.
Jeanne, R.L. (1970), Chemical defence of brood by a social wasp, *Science,* **168**, pp. 1465-6.
Jeannel, R. (1961), 'La Gondwanie et le peuplement de l'Afrique', *Annls. Mus. roy. Afr. cent.*, No. 102, pp. 1-161.
Johnson, H.M. (1953), 'Preliminary ecological studies of microclimates inhabited by the smaller arctic and subarctic mammals'. *Proc. 2nd Alaskan Sci. Conf.*, 1951, pp. 125-31.
Kachkarov, D.N. and Korovine, E.P. (1942), *La Vie dans les Deserts* (Ed. francaise par Th. Monod), Payot, Paris, 361 pp.
Kassas, M. (1970), 'Desertification versus potential for recovery in circum-Saharan territories'. *In* Dregne, H.E. (ed.), *Arid Lands in Transition, Amer. Assoc. Adv. Sci. Publ.*, Washington, D.C., No. 90, pp. 123-42.
Keast, A. (1968a), 'Evolution of mammals on southern continents. Introduction: the southern continents as backgrounds for mammalian evolution', *Quart. Rev. Biol.*, **43**, pp. 225-33.
Keast, A. (1968b), 'Evolution of mammals on southern continents. IV' Australian mammals: zoogeography and evolution', *Quart. Rev. Biol.*, **43**, pp. 373-451.
Keast, A. (1969), 'Evolution of mammals on southern continents. VII.

Comparisons of the contemporary mammalian faunas of the southern continents', *Quart. Rev. Biol.*, **44**, pp. 121-67.
Keast, A. (1971), 'Continental drift and the evolution of the biota on southern continents', *Quart. Rev. Biol.*, **46**, pp. 335-87.
Keast, J.A. and Marshall, A.J. (1954), 'The influence of drought and rainfall on reproduction in Australian desert birds', *Proc. Zool. Soc. Lond.*, **124**, pp. 493-9.
Keay, R.W.J. (1949), 'An example of Sudan zone vegetation in Nigeria', *J. Ecol.*, **37**, pp. 335-64.
Keilin, D. (1959), 'The problem of anabiosis or latent life: history and current concept', *Proc. Roy. Soc.* (B) **150**, pp. 149-91.
Kellas, L.M. (1955), 'Observations on the reproductive activity, measurement and growth rate of the dik-dik *(Rhynchotragus kerkii thomasi* Neumann)', *Proc. Zool. Soc. Lond.*, **124**, pp. 751-84.
Kemp, P.B. (1955), 'The termites of north-eastern Tanganyika: their distribution and biology', *Bull. Ent. Res.*, **46**, pp. 113-35.
Kevan. D.K.McE. (1962), *Soil Animals,* Witherby, London, 237 pp.
Kingdon, J. (1971), *East African Mammals,* Academic Press, London, Vol. 1, 446 pp.
Kirmiz, J.P. (1962), *Adaptation to Desert Environment, A Study of the Jerboa, Rat and Man,* Butterworth, London, 168 pp.
Koch, C. (1961), 'Some aspects of abundant life in the vegetationless sand of the Namib desert dunes', *J.S.W. Afr. Scient. Soc.*, **15**, pp. 8-34.
Koller, D. (1969), 'The physiology of dormancy and survival in desert environments', *Symp. Soc. Exp. Biol.*, **23**, pp. 449-69.
Koppen, W. (1931), *Grundriss der Klimakunde,* de Gruyter, Berlin, 384 pp.
Koskimies, J. (1955), 'Ultimate causes of cyclical fluctuations in numbers in animal populations', *Pop. Game-Res. Helsingf.*, No. 15, pp. 1-29.
Krogerus, R. (1932), 'Über die Ökologie und Verbreitung der Arthropoden der Triebsandgebiete an den Küsten Finnlands', *Acta zool. Fenn.* 12, pp. 1-308
Kühnelt, W. (1961), *Soil Biology* (Trans. N. Walker), Faber & Faber, London, 397 pp.
Lack, D. (1947), *'Darwin's Finches',* Univ. Press, Cambridge, 208 pp.
Lack, D. (1950), 'Breeding seasons in the Galapagos', *Ibis,* **92**, pp. 268-78.
Lack, D. (1954), *The Natural Regulation of Animal Numbers,* Clarendon Press, Oxford, 343 pp.
Lamprey, H.F. (1963), 'Ecological separation of the large mammal species in the Tarangire game reserve, Tanganyika', *E. Afr. Wildl. J.*, **1**, pp. 63-92.

Lawrence, R.F. (1953), *The Biology of the Cryptic Fauna of Forests,* Balkema, Cape Town, 408 pp.

Lawrence, R.F. (1959), 'The sand-dune fauna of the Namib desert', *S. Afr. J. Sci.,* **55**, pp. 233-9.

Laws, R.M. (1970), 'Elephants as agents of habitat and landscape change in Africa', *Oikos,* **21**, pp. 1-15.

Laws, R.M. and Clough, G. (1965), 'Observations on reproduction in the hippopotamus, *Hippopotamus amphibius* Linn', *Symp. Zool. Soc. Lond.,* No. 15, pp. 117-40.

Laws, R.M. and Parker, I.S.C. (1968), 'Recent studies on elephant populations in East Africa', *Symp. Zool. Soc. Lond.,* No. 21, pp. 319-59.

Laws, R.M. Parker, I.S.C. and Johnstone, R.C.B. (1970), 'Elephants and habitats in North Bunyoro, Ugando', *E. Afr. Wildl. J.,* **8**, pp. 163-80.

Lawson, G.W. (1966), *'Plant Life in West Africa',* Oxford Univ. Press, London, 150 pp.

Lees, A.D. (1955), *The Physiology of Diapause in Arthropods,* Univ. Press, Cambridge, 151 pp.

Legendre, R. and Cassagne-Mejean, F. (1967-68), 'Le problème de l'existence du continent gondwanien vu par des zoologistes (certitudes et incertitudes)', *Ann. Soc. Hort. Hist. Nat. Hérault.,* **107**, pp. 223-41; **108**, pp. 39-47; 109-117.

Leopold, A. (1933), *Game Management,* Scribner, New York, 481 pp.

Leopold, A. (1943), 'Deer irruptions', *Wis. Conserv. Dept. Publ.,* **321**, pp. 1-11.

Leopold, A.S. and Darling, F.F. (1953), *Wildlife in Alaska,* Ronald, New York, 129 pp.

Lindner, E. (1956), 'Zur Verbreitung der Dipteren (Zweiflügler) in den Hochregionen der Alpen', *Jb. Verein. Schutz. Alpenflanzen und Tiere, Munchen,* **1956**, pp. 121-8.

Livingstone, D.A. (1967), 'Post glacial vegetation of the Ruwenzori Mountains in equatorial Africa', *Ecol. Monogr.,* **37**, pp. 25-52.

Lockie, J.D. (1955), 'The breeding and feeding of jackdaws and rooks, with notes on carrion crows and other Corvidae', *Ibis,* **97**, pp. 341-69.

Lofts, B. and Murton, R.K. (1968), 'Photoperiodic and physiological adaptations regulating avian breeding cycles and their ecological significance', *J. Zool., Lond.,* **155**,pp. 327-94.

Lofts, B. Murton, R.K. and Westwood, N.J. (1967), 'The experimental demonstration of a post-nuptial refractory period in the turtle dove, *Streptopelia turtur', Ibis,* **109**, pp. 352-8.

Logan, R.F. (1968), 'Causes, climates, and distribution of deserts', *In* G.W. Brown jr. (ed.), *Desert Biology* **1**, pp. 21-50, Academic Press, New York.

Lousley, J.E. (1950), *Wild Flowers of Chalk and Limestone,* Collins, London, 254 pp.
Louw, G.N. Belonje, P.C. and Coetzee, H.J. (1969), 'Renal function, respiration, heart rate and thermoregulation in the ostrich *(Struthio camelus)', Sci. Pap. Namib Desert Res. Sta.,* No. 42, pp. 43-54.
Lull, R.S. (1940), *Organic Evolution* (revised ed.), Macmillan, New York, 743 pp.
McAtee, W.L. (1947), 'Torpidity in birds', *Amer. Midl. Nat.,* **38**, pp. 191-206.
McNab, B.K. (1971), 'On the ecological significance of Bergmann's rule', *Ecology,* **52**, pp. 845-54.
Macan, T.T. and Worthington, E.B. (1951), *Life in Lakes and Rivers,* Collins, London, 272 pp.
Macfadyen, A. (1963), *Animal Ecology. Aims and Methods* (2nd ed.), Pitman, London, 344 pp.
Maclean. G.L. (1967), 'The breeding biology and behaviour of the double-banded courser *Rhinoptilus africanus* (Temminck)', *Ibis,* **109**, pp. 556-69.
Mani, M.S. (1962), *Introduction to High Altitude Entomology,* Methuen, London, 302 pp.
Mani, M.S. (1968), *Ecology and Biogeography of High Altitude Insects,* Junk, The Hague, 527 pp.
Marshall, A.J. and Coombs, C.J.F. (1957), 'The interaction of environment, internal and behavioural factors in the rook *Corvus frugilegus* Linnaeus', *Proc. Zool. Soc. Lond.,* **128**, pp. 545-89.
Marshall, A.J. and Disney, H.J. de S. (1957), 'Experimental induction of the breeding season in a xerophilous bird', *Natur, Lond.,* **180**, pp. 647.
Marshall, F.H.A. (1942), 'Exteroceptive factors in sexual periodicity', *Biol. Rev.,* **17**, pp. 68-90.
Masefield, G.B. (1970), 'Food resources and production' *In* Garlick, J.P. Keay, and R.W.J. (eds.), 'Human ecology in the tropics', *Symp. Soc. Stud. Human Biology,* 9, pp. 59-66.
Matthews, L.H. and Carrington, R. (1970) (eds.), *The Living World of Animals,* Reader's Digest, London, 428 pp.
Medway, Lord (1972), 'Phenology of a tropical rain forest in Malaya', *Biol. J. Linn. Soc.,* **4**, pp. 117-46.
Mertens, R. (1948), *Die Tierwelt des tropischen Regenwaldes,* Kramer, Frankfurt am Main, 144 pp.
Miall, L.C. (1895), *The Natural History of Aquatic Insects,* Macmillan, London, 395 pp.
Michelmore, A.P.G. (1947), 'A popular misconception regarding humidity and the need for closer liaison between meteorologists and ecologists', *J. Ecol.,* **34**, pp. 107-10.

Miller, A.A. (1965), *Climatology* (9th ed. rep.) Methuen, London, 320 pp.

Moreau, R.E. (1950), 'The breeding seasons of African birds. Parts 1 and 1', *Ibis,* **92,** pp. 223-67; 419-33.

Moreau, R.E. (1961), 'Problems of Mediterranean-Saharan migration', *Ibis,* **103,** pp. 373-427.

Moreau, R.E. (1966), *'The Bird Faunas of Africa and its Islands',* Academic Press, London, 424 pp.

Moreau, R.E. (1967), 'Water birds over Sahara'. *Ibis,* **109,** pp. 323-59.

Moreau, R.E. (1969), 'Climatic changes and the distribution of forest vertebrates in West Africa', *J. Zool., Lond.,* **158,** pp. 39-61.

Mossman, A.S. (1955), 'Light penetration in relation to small mammal abundance', *J. Mammal.,* **36,** pp. 564-6.

Mueller-Dombois, D. (1971), 'Crown distortion and elephant distribution in the woody vegetations of Ruhuna National Park, Ceylon', *Ecology,* **53,** pp. 208-26.

Mutere, F.A. (1965), 'Delayed implantation in an equatorial fruit bat', *Nature, Lond.,* **207,** p. 780.

Myers, J.G. (1935), 'Nesting associations of birds with social insects', *Trans. R. Ent. Soc. Lond.,* **83,** pp. 11-22.

Newbigin, M. (1936), *Plant and Animal Geography,* Methuen, London, 298 pp.

Noble, G.K. (1931), *The Biology of the Amphibia,* McGraw-Hill, New York, 577 pp.

Norman, J.R. (1963), *A History of Fishes* (2nd ed. by P.H. Greenwood), Benn, London, 398 pp.

Odum, E.P. (1971), *Fundamentals of Ecology* (3rd ed.), Saunders, Philadelphia, 574 pp.

Olaniyan, C.I.O. (1968), *An Introduction to West African Animal Ecology,* Heinemann, London, 167 pp.

Olson, E.C. (1972), 'The habitat: climatic change and its influence on life and habitat'. *In* T-W-Fiennes, R.N. (ed.), *Biology of Nutrition,* Pergamon, Oxford, pp. 267-305.

Orr, R.T. (1945), 'A study of captive finches of the genus *Geospiza*' *Condor,* **47,** pp. 177-201.

Owen, D.F. (1966), *Animal Ecology in Tropical Africa,* Oliver & Boyd, Edinburgh, 122 pp.

Owen, D.F. (1971), *Tropical Butterflies,* Clarendon Press, Oxford, 214 pp.

Pantin, C.F.A. (1964), 'Homeostasis and the environment', *Symp. Soc. Exp. Biol.,* **18,** pp. 1-6.

Pantin, C.F.A. (1968), *The Relations Between the Sciences,* Univ. Press, Cambridge, 206 pp.

Paris, O.H. (1963), 'The ecology of *Armadillidium vulgare* (Isopoda:

Oniscoidea) in California grassland: food, enemies and weather', *Ecol. Monogr.*, **33**, pp. 1-22.

Patterson, B. and Pascual, R. (1968), 'Evolution of mammals on southern continents. V. The fossil mammal fauna of South America', *Quart. Rev. Biol.*, **43**, pp. 409-51.

Pearson, P.P. (1954), 'Habits of the lizard, *Liolaemus multiformis multiformis* at high altitudes in southern Peru', *Copeia*, **1954**, pp. 111-16.

Percival, A.B. (1928), *A Game Ranger on Safari*, Nisbet, London, 305 pp.

Perry, J.S. (1953), 'The reproduction of the African elephant, *Loxodonta africana*', *Phil. Trans. Roy. Soc.*, (B) **237**, pp. 93-149.

Pierre, F. (1958), *Écologie et peuplement entomologique des sables vifs du Sahara nord-occidental*, Centre nat. Recherche scientifique, Paris, 332 pp.

Platt, A.P. and Brower, L.P. (1968), 'Mimetic versus disruptive coloration in intergrading populations of *Limenitis arthemis* and *astyanax* butterflies', *Evolution*, **22**, pp. 699-718.

Polunin, N. (1960), *Introduction to Plant Geography*, Longmans, London, 640 pp.

Pond, A.W. (1962), *The Desert World*, Nelson, New York, 342 pp.

Porsild, A.E. Harington, C.R. and Mulligan, G.A. (1967), '*Lupinus arcticus* Wats. grown from seeds of Pleistocene age', *Science*, **158**, pp. 113-4.

Poulton, E.B. (1908), *Essays on Evolution*, Clarendon Press, Oxford, 479 pp.

Prakash, I. (1959), 'Hypertrophy of the bullae tympanicae in the desert mammals', *Sci. Culture*, **24**, pp. 580-2.

Prakash, I. (1960), 'Breeding of mammals in the Rajasthan desert, India'. *J. Mammal.*, **41**, pp. 386-9.

Rahm, U. (1970), 'Note sur la reproduction des Sciurides et Murides dans la forêt equatoriale au Congo', *Revue Suisse Zool.*, 77, pp. 635-46.

Rainey, R.C. (1951), 'Weather and the movements of locust swarms: a new hypothesis', *Nature, Lond.*, **168**, pp. 1057-60.

Reichle, D.E. (1970) (ed.), *Analysis of Temperate Forest Ecosystems*, Chapman & Hall, London, 304 pp.

Reid, G.K. (1961), *Ecology of Inland Waters and Estuaries*, Reinhold, New York, 375 pp.

Reinig, W.F. (1932), 'Beiträge zur Faunistik des Pamir Gebietes', *Wiss, Ergeb. Alai-Pamir-Expedition* 1928, 1 (3) Ökologie und Tiergeographie, pp. 1-195.

Reitan, C.R. and Green, C.R. (1968), 'Weather and climate of desert environments', *In McGinnies, W.G., Goldman, B.J. and Paylore, P.*

(eds.), *Deserts of the World,* Univ. Press, Arizona, pp. 19-92.
Rettenmeyer, C.W. (1961), 'Observations on the biology and taxonomy of flies found over swarm raids of army ants (Diptera: Tachinidae, Conopidae)', *Univ. Kansas Sci. Bull.*, **42**, pp. 993-1066.
Rettenmeyer, C.W. (1963), 'Behavioral studies of army ants.' *Univ. Kansas Sci. Bull.*, **44**, pp. 281-465.
Rettenmeyer, C.W. (1970), 'Insect mimicry', *Ann. Rev. Ent.*, **15**, pp. 43-74.
Reynolds, V. (1965), *Budongo. An African Forest and its Chimpanzees,* Natural History Press, New York, 253 pp.
Richards, P.W. (1952), *The Tropical Rain Forest. An Ecological Study,* Univ. Press, Cambridge, 450 pp.
Richards, P.W. (1970), *The Life of the Jungle,* McGraw-Hill, New York, 232 pp.
Robinson, M.H. (1969), 'Defenses against visually hunting predators', *Evol. Biol.*, **3**, pp. 225-59.
Roe, F.G. (1951), *The North American Buffalo: a Critical Study of the Species in its wild State,* Univ. Toronto Press, Toronto, 957 pp.
Rowan, W. (1931), *The Riddle of Migration,* Williams & Wilkins, Baltimore, 151 pp.
Rowan, W. (1932), 'Experiments in bird migration. III. The effects of artificial light, castration and certain extracts on the autumn movement of the American crow *(Corvus brachyrhynchos)', Proc. Nat. Acad. Sci. U.S.A.*, **18**, pp. 639-54.
Russell, E.J. (1957), *The World of the Soil,* Collins, London, 237 pp.
Ruttner, F. (1953), *Fundamentals of Limnology* (Trans. D.G. Frey and F.E.J. Fry), Univ. Toronto Press, Toronto, 242 pp.
Rzóska, J. (1961), 'Observations on tropical rainpools and general remarks on temporary waters', *Hydrobiol.*, **17**, pp. 265-86.
Salisbury, E. (1952), *Downs and Dunes. Their Plant Life and its Environment,* Bell, London, 328 pp.
Salt, G. (1954), 'A contribution to the ecology of Upper Kilimanjaro', *J. Ecol.*, **42**, pp. 275-423.
Salt, R.W. (1961), 'Principles of insect cold-hardiness', *Ann. Rev. Ent.*, **6**, pp. 55-74.
Salt, R.W. (1964), 'Terrestrial animals in cold: Arthropods'. *In* Dill, D.B. (ed.), *Handbook of Physiology,* Sect. 4, pp. 349-55, Amer. Physiol. Soc., Washington, D.C.
Salt, R.W. (1969), 'The survival of insects at low temperatures', *Symp. Soc. Exp. Biol.*, **23**, pp. 331-50.
Sankey, J. (1966), *Chalkland Ecology,* Heinemann, London, 137 pp.
Savory, T.H. (1971), *Biology of the Cryptozoa,* Merrow, Watford, Herts., 48 pp.
Schmidt-Nielsen, B. and Schmidt-Nielsen, K. (1950), 'Evaporative water

loss in desert rodents in their natural habitat', *Ecology,* **31**, pp. 75-85.
Schmidt-Nielsen, B. Schmidt-Nielsen, K. Houpt. T.R. and Jarnum, S.A. (1956), 'Water balance of the camel', *Amer. J. Physiol.,* **185**, pp. 185-94.
Schmidt-Nielsen, K. (1962), 'Comparative physiology of desert mammals', *Univ. Missouri Agri. Exp. Sta.,* Special Report No. 21, pp. 1-31.
Schmidt-Nielsen, K. (1964), *Desert Animals. Physiological Problems of Heat and Water,* Clarendon Press, Oxford, 277 pp.
Schmidt-Nielsen, K. (1972), *How Animals Work,* Univ. Press, Cambridge, 114 pp.
Schmidt-Nielsen, K. Kanwisher, J. Lasiewski, R.C. Cohn, J.E. and Bretz, W.L. (1969), 'Temperature regulation and respiration in the ostrich', *Condor,* **71**, pp. 341-52.
Schmidt-Nielsen, K. and Schmidt-Nielsen, B. (1950), 'Do kangaroo rats thrive when drinking sea water?', *Amer. J. Physiol.,* **160**, pp. 291-4.
Schmidt-Nielsen, K. and Schmidt-Nielsen, B. (1952), 'Water metabolism of desert mammals', *Physiol. Rev.,* **32**, pp. 135-66.
Schmidt-Nielsen, K. Schmidt-Nielsen, B. Houpt, T.R. and Jarnum, S.A. (1956), 'The question of water storage in the stomach of the camel', *Mammalia,* **20**, pp. 1-15.
Schneirla, T.C. (1971), *Army Ants. A Study in Social Organisation.* Freeman, San Francisco, 349 pp.
Sclater, P.L. (1858), 'On the general geographic distribution of the members of the class Aves', *J. Linnean Soc. Lond., Zool.,* **2**, pp. 130-45.
Scott, H. (1912), 'A contribution to the knowledge of the fauna of Bromeliaceae', *Ann. Mag. Nat. Hist.,* (8) **10**, pp. 424-38.
Scott, H. (1952), 'Journey to the Gughé highlands (Southern Ethiopia), 1948-9; biogeographical research at high altitudes', *Proc. Linn. Soc. Lond.,* **163**, pp. 85-189.
Scott, H. (1958), 'Biogeographical research in high Simien (Northern Ethiopia), 1952-53', *Proc. Linn. Soc. Lond.,* **170**, pp. 1-91.
Séguy, E. (1950), 'La biologie des Diptères', *Encycl. Ent.,* (A) **26**, pp. 1-609.
Seton, E.T. (1909), *Life Histories of Northern Animals: an Account of the Mammals of Manitoba,* Scribner, New York, 2 vols., 1220 pp.
Shaller, F. (1968), *Soil Animals,* Univ. Michigan Press, Ann Arbor, 144 pp.
Shantz, H.L. and Zon, R. (1924), 'Natural vegetation', *U.S. Dept. Agric. Atlas of American Agriculture,* **1** (E), pp. 1-29.
Shaukat, S.S. and Quadir, S.A. (1971), 'Multivariate analysis of the vegetation of calcareous hills around Karachi', *Vegetatio,* **23**, pp. 235-53.

Shelford, V.E. and Olson, S. (1935), 'Sere, climax and influent animals with special reference to the transcontinental coniferous forest of North America', *Ecology,* **16**, pp. 375-402.
Sidorowicz, J. (1971), 'Zoogeographical regionalization of the world based on the distribution of the members of the Order Carnivora (Mammalia)', *Acta Zool. Cracoviensia,* **16**, pp. 309-95.
Siivonen, L. and Koskimies, J. (1955), 'Population fluctuations and the lunar cycle', *Pap. Game Res. Helsingf.,* No. 14, pp. 1-10.
Sikes, S.K. (1966), 'The African elephant: the background to its present-day ecological status in the Murchison (Uganda) and Tsavo (Kenya) National Parks and environs', *Rev. Zool. Bot. Afr.,* **74,** pp. 255-72.
Sikes, S.K. (1971), *The Natural History of the African Elephant,* Weidenfeld & Nicolson, London, 397 pp.
Sinclair, J.G. (1922), 'Temperatures of the soil and air in a desert', *Mon. Weather Rev.,* **50,** pp. 142-4.
Smith, A. (1970), *The Seasons, Rhythms of Life: Cycles of Change,* Weidenfeld & Nicolson, London 318 pp.
Smith, A.P. (1972), 'Buttressing of tropical trees: a descriptive model and new hypothesis', *Amer. Nat.,* **106,** pp. 32-46.
Smith, K.D. (1955), 'The winter breeding season of land birds in Eastern Eritrea', *Ibis,* **97,** pp. 480-507.
Smith, R.F. (1954), 'The importance of the microenvironment in insect ecology', *J. Econ. Ent.,* **47,** pp. 205-10.
Søfrensen, T. (1941), 'Temperature relations and phenology of the northeast Greenland flowering plants', *Medd. Grønland,* **125** (9) pp. 1-305.
Southern, H.N. and Hook. O. (1963a), 'A note on small mammals in East African Forests', *J. Mammal,* **44,** pp. 126-9.
Southern, H.N. and Hook. O. (1963b), 'Notes on breeding of small mammals in Uganda and Kenya', *Proc. Zool. Soc. Lond.,* **140,** pp. 503-15.
Southwood, T.R.E. (1962), 'Migration of terrestrial arthropods in relation to habitat', *Biol. Rev.,* **37,** pp. 171-214.
Spedding, L.W.R. (1971), *Grassland Ecology,* Clarendon, Oxford, 221 pp.
Spellerberg, I.F. (1972a), 'Thermal ecology of allopatric lizards *(Sphenomorphus)* in Southeast Australia. The environment and lizard critical temperatures', *Oecologia (Berl.),* **9,** pp. 371-83.
Spellerberg, I.F. (1972b), *Idem.* II. 'Physiological aspects of thermoregulation', *Oecologia (Berl.),* **9,** pp. 385-98.
Spellerberg, I.F. (1972c), *Idem.* III. 'Behavioural aspects of thermoregulation', *Oecologia (Berl.),* **11,** pp. 1-16.
Spurr, S.H. (1964), *Forest Ecology,* Ronald Press, New York, 352 pp.

Stebbing, E.P. (1954), Forests, Aridity and Deserts. *In* Cloudsley-Thompson, J.L. (ed.), *Biology of Deserts,* Inst. Biol., London, pp. 123-8.
Suess, E. (1904), *The Face of the Earth (Das Antlitz der Erde)* (Transl. H.B.C. Sollas), Clarendon Press, Oxford, **1**, 604 pp.
Sukachev, V. and Dylis, D. (1968), *Fundamentals of Forest Biogeocoenology* (Trans. J.M.D. Maclennan), Oliver & Boyd, Edinburgh, 672 pp.
Sumner, F.B. (1921), 'Desert and lava-dwelling mice, and the problem of protective coloration in mammals', *J. Mammal,* **2**, pp. 75-86.
Supan, A. (1896), *Grundzuge der physischen Erdkunde,* de Gruyter, Leipzig, 706 pp.
Swan, L.W. (1961), 'The ecology of the High Himalayas', *Sci. Amer.,* **205** (4), pp. 68-78.
Talling, J.F. and Rzóska, J. (1967), 'The development of plankton in relation to hydrological regime in the Blue Nile', *J. Ecol.,* **55**, pp. 637-62.
Tansley, A. (1939), *The British Islands and their Vegetation,* Univ. Press, Cambridge, 930 pp.
Tansley, A.G. (1949), *Britain's Green Mantle Past, Present and Future,* Allen & Unwin, London, 294 pp.
Tarling, D.H. (1972), 'Another Gondwanaland', *Nature, Lond.,* **238**, pp. 92-3.
Tercafs, R.R. (1962), 'Observations ecologiques dans le massif du Tibesti (Tchad)', *Rev. Zool. Bot. Afr.,* **66**, pp. 107-26.
Tevis, L. and Newell, I.M. (1962), 'Studies on the biology and seasonal cycle of the red velvet mite *Dinothrombium pandorae* (Acari, Trombidiidae)', *Ecology,* **43**, pp. 497-505.
Thiele, H.V. and Kirchner, H. (1958), 'Über die Körpergrosse der Gebirgs- und Flachlandpopulationen einiger Laufkäfer (Carabidae)', *Bonn. Zool. Beitr.,* **9**, pp. 294-302.
Thomson, A. Landsborough (1926), *Problems of Bird Migration,* Witherby, London, 350 pp.
Thomson, A. Landsborough (1949), *Bird Migration: a Short Account* (3rd ed.), Witherby, London, 183 pp.
Thornton, D.D. (1971), 'The effect of complete removal of hippopotamus on grassland in the Queen Elizabeth National Park, Uganda', *E. Afr. Wildl. J.,* **9**, pp. 47-55.
Thorpe, W.H. (1942), 'Observations on *Stomoxys ochrosoma* Speiser (Diptera, Museidae) as an associate of army ants (Dorylinae) in East Africa', *Proc. R. Ent. Soc. Lond.* (A) **17**, pp. 38-41.
Thorpe, W.H. (1950), 'Plastron respiration in aquatic insects', *Biol. Rev.* **25**, pp. 344-90.
Thorpe, W.H. and Crisp, D.J. (1947), 'Studies on plastron respiration.

1. The biology of *Aphelocheirus* (Hemiptera, Aphelocheiridae (Naucoridae)) and the mechanism of plastron retention', *J. Exp. Biol.,* **24,** pp. 227-69.

Tinbergen, N. (1963), 'On aims and methods in ethology', *Z. Tierpsychol.,* **20,** pp. 410-33.

Treshow, M. (1970), *Environment and Plant Response,* McGraw-Hill, New York, 422 pp.

Trewartha, G.T. (1968), *An Introduction to Climate,* McGraw-Hill, New York, 408 pp.

Twomey, A.C. (1945), 'The bird population of an elm-maple forest with special reference to aspection, territorialism, and coactions', *Ecol. Monogr.,* **15,** pp. 173-205.

Ule, E. (1904), 'Ameisengärten im Amazonas Gebeit', *Engler's Bot. Jahrb.,* **30,** pp. 45-52.

Ullyott, P. (1939), 'Die Taglichen Wanderungen der planktonischen Susswasser-Crustaceen', *Int. Rev. Hydrobiol.,* **38,** pp. 262-84.

Urquhart, F.A. (1960), *The Monarch Butterfly,* Univ. Press, Toronto, 361 pp.

Uvarov, B.P. (1931), 'Insects and climate', *Trans. Ent. Soc. Lond.,* **79,** pp. 1-247.

Vaartaja, O. (1949), 'High surface soil temperatures. On methods of investigation, and thermocouple observations on a wooded heath in the south of Finland', *Oikos,* **1,** pp. 6-28.

Vesey-Fitzgerald, D.F. (1963), 'Central African grasslands', *J. Ecol.,* **51,** pp. 243-74.

Viitanen, P. (1967), 'Hibernation and seasonal movements of the viper, *Vipera berus berus* (L.) in southern Finland', *Ann. Zool. Fenn.,* **4,** pp. 472-546.

Wagner, H.O. (1959), 'Nestplatzwahl und den Nestbau auslösende Reise bei einigen mexikanischen Vogelarten', *Z. Tierpsychol.,* **16,** pp. 297-301.

Wallace, A.R. (1860), 'On the zoological geography of the Malay Archipelago', *J. Linn. Soc. (Zool.),* **4,** pp. 172-84.

Wallace, A.R. (1869), *The Malay Archipelago,* Macmillan, London, 638 pp.

Wallace, A.R. (1876), *Geographical Distribution of Animals,* Macmillan, London, **1,** 503 pp.; **2,** 607 pp.

Wallén, C.C. (1966), 'Arid zone meteorology'. *In* E.S. Hill (ed.), *Arid Lands. A Geographical Appraisal,* Methuen, London, pp. 31-52.

Wallwork, J.A. (1970), *Ecology of Soil Animals,* McGraw-Hill, London, 283 pp.

Wallwork, J.A. (1973), 'Zoogeography of some terrestrial micro-arthropoda in Antarctica', *Biol. Rev.,* **48,** pp. 233-59.

Ward, H.B. and Whipple, G.C. (1959), *Fresh-water Biology* (ed. W.T.

Edmondson) (2nd ed.), Wiley, New York, 1248 pp.
Wasmund, E. (1934), 'Die physiologische Bedeutung des limnischen Hydroklimas', *Arch. Hydrobiol.*, **27**, pp. 162-98.
Watts, I.E.M. (1955), *Equatorial Weather with Particular Reference to South East Asia,* Univ. Press, London, 223 pp.
Wegener, A.L. (1924), *The Origin of Continents and Oceans,* (3rd ed. trans. by J.G.A. Skerl), Methuen, London, 212 pp.
Weiner, J.S. (1971), *Man's Natural History,* Weidenfeld & Nicholson, London, 255 pp.
Weir, J. and Davidson, E. (1965), 'Daily occurrence of African game animals at water holes during dry weather', *Zool. Africana,* **1**, pp. 353-68.
Welch, P.S. (1935), *Limnology,* McGraw-Hill, New York, 471 pp.
Wells, H.G., Huxley, J. and Wells, G.P. (1931), *The Science of Life,* Cassell, London, 896 pp.
Wesenberg-Lund, C. (1911), 'Grundzüge der Biologie und Geographie des Süsswasserplanktons, nebst Bemerkungen über Hauptprobleme zukünftiger limnologischer Forschungen. (Aus. d. Dän. übers.)', *Int. Rev. Hydrobiol.,* **3** *(Suppl.)* pp. 1-44.
Wesenberg-Lund, C. (1930), 'Contributions to the biology of the Rotifera. Part II. The periodicity and sexual periods', *D. Kgl. Danske Vidensk. (Nat. Mak.),* **9**, (11), pp. 1-230.
Wheeler, W.M. (1910), *Ants. Their Structure and Development,* Columbia Univ. Press, New York, 663 pp.
Wieser, W. (1963), 'Die Bedeutung der Tageslange fur das Einsetzen der Fortpflanzungsperiode bei *Porcellio scaber* Latr. (Isopoda)', *Z. Naturforsch.,* **18**, pp. 1090-2.
Wieser, W. Schweizer, G. and Hortenstein, R. (1969), Patterns in the release of gaseous ammonia by terrestrial isopods', *Oecologia (Berl.),* **3**, pp. 390-400.
Wigglesworth, V.B. (1964), *The Life of Insects,* Weidenfeld & Nicolson, London, 360 pp.
Williams, C.B. (1930), *The Migration of Butterflies,* Oliver & Boyd, Edinburgh, 473 pp.
Williams, C.B. (1958), *Insect Migration,* Collins, London, 235 pp.
Williams, C.B. (1954), 'Some bioclimatic observations in the Egyptian desert'. *In* Cloudsley-Thompson, J.L. (ed.), *Biology of Deserts,* Inst. Biol., London, pp. 18-24.
Willoughby, E.J. (1969), 'Desert coloration in birds of the central Namib desert', *Sci. Pap. Namib Desert Res. Sta.* No. 44, pp. 59-68.
Wynne-Edwards, V.C. (1962), *Animal Dispersion in Relation to Social Behaviour,* Oliver & Boyd, Edinburgh, 653 pp.
Yapp, W.B. (1953), 'The high-level woodlands of the English Lake District', *North West Nat.,* **24**, pp. 190-207; 370-83.

Young, J.Z. (1950), *The Life of Vertebrates*, Clarendon Press, Oxford, 767 pp.

APPENDIX 1

Classification of World Climates and Vegetation

A. *Hot climates:* Mean annual temperatures above 21°C (70°F).

1. Equatorial	(Rain-forest)
2. Tropical marine	(Rain-forest)
3. Tropical continental	(Savannah)

B. *Warm temperate (or sub-tropical):* No month below 6°C (43°F).

1. Western margin	(Temperate forest)
2. Eastern margin	(Temperate forest)
3. Continental	(Steppe)

C. *Cool temperate:* One to five months below 6°C (43°F).

1. Marine	(Temperate forest)
2. Continental	(Taiga or steppe)

D. *Cold climates:* Six or more months below 6°C (43°F).

1. Marine	(Taiga)
2. Continental or boreal	(Taiga or steppe)

E. *Arctic climates:* No month above 10°C (50°F).
(Tundra)

F. *Desert climates:* Low rainfall.

1. Hot: No month below 6°C (43°F)
2. Cold: One or more months below 6°C (43°F)

G. *Mountain climates:*

APPENDIX 2

The Deserts of the World

	Square kilometres (millions)	Square miles (millions)
Sahara Desert	9.1	3.5
Australian Desert	3.4	1.3
Arabian Desert	2.6	1.0
Turkestan Desert	1.9	0.75
Great American Desert (includes the *Great Basin, Mojave, Sonoran* and *Chihuahuan Deserts* of south western North America)	1.3	0.5
Patagonian Desert (Argentina)	0.67	0.26
Thar Desert (India)	0.60	0.23
Kalahari and *Namib Deserts* (South West Africa)	0.57	0.22
Takla Makan Desert including the *Gobi Desert* (western China to Mongolia)	0.52	0.20
Iranian Desert (Persia)	0.39	0.15
Atacama Desert (Peru and Chile) (This has the lowest precipitation of all, averaging under 2.5mm (1 in.) of rain per year.)	0.36	0.14

INDEX TO AUTHORS CITED

GENERAL INDEX

Index of French words

This index lists some of the French words you might see on signs and notices. Look up the page reference to find out what they mean.